Charles Seale-Hayne Library
University of Plymouth
(01752) 588 588
LibraryandITenquiries@plymouth.ac.uk

From Idea to Working Model

Wiley Series on Systems Engineering and Analysis
HAROLD CHESTNUT, Editor

Chestnut
Systems Engineering Tools

Wilson and Wilson
Information, Computers, and System Design

Hahn and Shapiro
Statistical Models in Engineering

Chestnut
Systems Engineering Methods

Rudwick
*Systems Analysis for Effective Planning:
Principles and Cases*

Wilson & Wilson
From Idea to Working Model

From Idea to Working Model

Ira G. Wilson/Marthann E. Wilson

John Wiley & Sons, Inc., New York · London · Sydney · Toronto

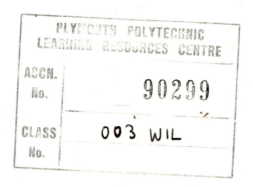
Copyright © 1970, by John Wiley & Sons, Inc.

Library of Congress Catalogue Card Number: 70-105388

SBN 471 95198 6

Printed in the United States of America

10 9 8 7 6 5 4 3 2 1

to our growing i. g.

SYSTEMS ENGINEERING AND ANALYSIS SERIES

In a society which is producing more people, more materials, more things, and more information than ever before, systems engineering is indispensable in meeting the challenge of complexity. This series of books is an attempt to bring together in a complementary as well as unified fashion the many specialties of the subject, such as modeling and simulation, computing, control, probability and statistics, optimization, reliability, and economics, and to emphasize the interrelationship between them.

The aim is to make the series as comprehensive as possible without dwelling on the myriad details of each specialty and at the same time to provide a broad basic framework on which to build these details. The design of these books will be fundamental in nature to meet the needs of students and engineers and to insure they remain of lasting interest and importance.

Preface

The innovative process furnishes the motive power that maintains progress in our modern society. It is responsible for the great changes and advances characteristic of our age. It also gives us our works of art, our musical productions, and our literature. In fact, all of the advances taking place in our civilized world depend on the innovative process.

Because of its all-pervasive importance, we all need to understand its nature. Only with this understanding can we hope to improve our abilities to innovate.

The procedure presented in this book provides a list of the necessary steps for any innovative process. The list is intended to be complete.

Theory shows that these steps must be arranged in a particular pattern. There is a first step and a last step. The steps between follow in a particular order.

In describing the procedure, we emphasize that as new information becomes available it may be necessary or desirable to return to an earlier step in the pattern and to make revisions. These operations are often called "recycling." Also it is emphasized that one does not always go through the complete list of steps. Although many ideas are conceived, few are carried through to completion. The reasons are explained.

A step-by-step procedure for innovation can help you to systematize your work. When used by a supervisor or manager, it points out the facts to be taken into account when evaluating and checking progress toward the wanted goals. All of us are innovators at one time or another; many of us work full time at innovation.

Worthwhile innovation always starts with a good "idea" of some need or want that can or should be fulfilled. The idea may or may not be expressed in words. Hopefully the end result of the innovative process is something new or different that fills the need. Notice particularly that the result is *wanted*.

Innovation is the result of mental work done by the human brain. Oh, of course, a computer can help a bit here and there on some innovative projects, but by and large, a computer is only a tool—a powerful and versatile tool. Nevertheless it is *only* a tool. Computers have both built-in and inherent limitations, which must be overcome in order to increase the amount of help they can be in innovation.

What does the human brain do in innovation?

Three things. It gathers information. It processes this information. It makes decisions.

An understanding of the three processes should be based on theory. To introduce you to such theory is one of the purposes of this book. To give you a practical and useful procedure for using the theory (even without understanding it completely) is a second—and very important—purpose.

By itself theory does not necessarily provide enough background to develop and understand the innovative process. So what do you do? You study examples of the innovative process, and, if the examples are carefully chosen, then by inductive reasoning you can work out a general procedure. This is the approach that we have taken here.

Where can you find examples to study? In the literature that describes how complex systems such as rocket systems, computers, and communication systems have been developed.

The study of these system projects brings out the common elements in innovative undertakings. Thus, with theory to furnish guidelines and with the patterns gleaned from system design, the necessary procedures for successful innovation emerge.

Both theory and practice have been used to develop the procedure described, and even a beginner in a field of innovation or design can understand how it works. Once you know how it works, you will be well on your way toward applying it. Also, those with experience in creating new ideas and things can observe ways to improve their efforts.

The required kinds of information and the permissible operations on the information are given for each step in the pattern.

Particular techniques and tools are needed; hence many practical suggestions are included, and, even more, useful checklists will help you with your own particular problems.

This book presents a soundly conceived and well-thought-out procedure. It is based on our interest in and experience with the innovative process. It is basically more theoretical, sounder in reasoning, and more complete than any other exposition we know about.

Because of the importance of innovation in our world, a long list of books has been written about it. Some are a few hundred years old.

Many have titles including the words "innovation," "thinking," or "creativity." Some of these books present a serious attempt to understand particular aspects of the innovative process, but many use a how-to-do-it approach with little or no theoretical basis. Many have essentially the same list of recommended steps for solving a problem. Unfortunately the recommendations are usually vague at best—and incomplete at worst. In most of the available literature emphasis is placed on particular aspects of the problem. The whole picture is not presented in a unified way.

The present volume provides the necessary background for each step. There is little or no mathematics per se. Some mathematical background may be helpful, but it is not necessary for an understanding of the subject matter. There are many examples to help clarify each point as it is made.

You will not be surprised to learn that this book depends on information gathered from many different disciplines. Neither will you be surprised to find that it coordinates some of the known results. However, it brings out many other points seldom, if ever, explicitly stated. Above all, it takes a different point of view from any of the previous works that we have consulted.

Because innovation is based on information processing, a body of knowledge built up over the last decades can be tapped: information theory. This very general theory has found many applications in various branches of knowledge. As far as we know, however, it has had little application to the study of innovation.

Information theory, as it has developed, is based on precise statements. It has been evolved by mathematicians. Some of its basic theorems can provide guidelines to separate the possible from the impossible in information processing. In particular, the possible kinds of changes of a set of items of information can be tabulated. Thus you can learn all of the possible approaches to be used in thinking out an innovation problem.

Since innovation results from mental effort, quite naturally questions about the human brain arise: for instance, what effects do training, experience, and age have on innovative ability? These questions are discussed, as are questions about work habits: what are the ways to minimize a drawback such as the limited memory capacity of the brain?

Many examples are used to illustrate particular points. In the final chapter the procedure is applied to put together a functional model of the brain. To do this the procedure and techniques are used. You can see how they work, and you may also learn some of the difficulties encountered in creating such models.

Planners, designers, and engineers whose work is intended to provide innovation certainly should be aware of and employ the most efficient procedures. It matters not whether your job is electrical, mechanical, chemical, or something else. The steps are the same; and to do a good job you should be familiar with, understand, and apply the best techniques available.

Working hand in hand with designers to produce the results of innovation are the "managers." In a large company the lower level managers supervise and work closely with the designers. Managers in the upper echelons are responsible for providing facilities and money. They make the "go–no go" decisions about projects proposed and under way. Top managers monitor the progress and reward or withhold rewards, depending on performance. In doing so they are responsible for the success of their talented innovators.

Some large laboratories use a shotgun approach to innovation: they hire hundreds, thousands, or tens of thousands of people and expect a generous return on their investment. The facts show, however, that only a small percentage of the people in large organizations get most of the patents and write most of the papers in learned publications. Therefore, if the application of the procedures of innovation increased the productivity by only a modest amount, it would be well worthwhile. For this reason this book discusses the characteristics to look for in selecting prospective innovators and how to avoid putting roadblocks in their way.

Now to another topic: training. Educators have as one of their missions the training of the innovations of the future. As part of this mission they should know how to recognize innovative talent at an early age and should nurture talent when it is found. Unfortunately our modern educational system tends to work against or suppress innovative talent. Some of the ways in which this occurs are discussed. Also, differences between our present methods of teaching with examinations are contrasted with real-life situations in which there is no teacher and no answer until you reach your objective.

Psychologists and behavioral scientists—even our psychiatrists—are all interested in one aspect or another of the innovative process. Group meetings of psychologists and behavioral scientists have had numerous round-table discussions of the many important factors involved; for example, they have studied the types of people who have been innovators in the past. One of the objectives is to identify and select the few talented people so that they may be given every advantage.

Physiologists concern themselves with the brain and how it works.

Working with the psychologists, they have created block-diagram models which are intended to account for some of the observed facts. An understanding of how the brain gathers and processes information is central to their basic studies.

Innovation is usually the work of a single man or a small team (even though they may be members of a large research and development laboratory). Yet the interaction between individuals and their peers, superiors, and other workers is important to success. Therefore these aspects of innovation are discussed in considerable detail.

In summary, the effects of innovation—or its lack—are important to economists as a contributing factor to our national planning and budget.

In performing the necessary steps in the innovation process you spend time, money, or (more likely) both. Now, in making the parts for an automobile and assembling them, time and money are expended. Hence the value of the completed car is greater than that of the raw materials. Economists call the increase the "increase in value added by manufacture." Similarly, in gathering and processing the information which hopefully will lead to an innovation, the amount and value of the available information are increased. By analogy an economist might call the increase "the increase in value added by innovation." This idea was contributed by a reviewer of the manuscript and should be pursued; for example, an input–output study of information growth might be made similar to the input–output model of the United States (and other economic units). Such a model would open up a whole new field to explore.

* * *

The interest of one of us (I.G.W.) in the problems of efficient innovation goes back several decades. This interest was aroused in the course of my participation in the planning, research, and development of numerous large technological systems. As a result, I have collected a large amount of pertinent literature over the years. This collection has been read carefully, pondered, and searched for relevant ideas. Unfortunately, as mentioned before, there is far more chaff than wheat.

One item in my collection, which I have kept for a number of years, is an unpublished paper by a member of the Bell Telephone Laboratories technical staff: Estill I. Green. In his own delightful way he summarized much of the existing literature at the time in a modest dozen pages. I am grateful for the privilege of having access to his work.

Oscar Myers gave a course on creativity and invention at the Bell laboratories in 1958–1959. I made numerous comments on his early drafts, and some of the ideas in his notes have been incorporated into this book.

*   *   *

In my doctoral dissertation, my work at Teachers College of Columbia University, and in private practice I (M.E.W.) have also explored new fields in the areas of physiology and relaxation.

*   *   *

A reviewer unknown to us went over the manscript with great care and made many helpful comments. All of them were acted on. We think a better book has resulted and hereby acknowledge his contribution.

Last, but not least, we acknowledge the work of Mrs. Ruth Shuping who took the original draft of the manuscript off the tapes and typed the several revisions with great care and enthusiasm.

Ira G. Wilson

Marthann E. Wilson

Tucson, Arizona
February 1970

Contents

Chapter 9   Information Processing by the Brain 211

From Idea to Working Model

1

Introduction

"Ideas" and "Thinking"

This is a book about *ideas* and *thinking*.

We shall discuss two quite different meanings of the word "idea." In one use of the word, we shall mean an "idea" for a new project. We call this an "idea 1." In the other use, we shall mean an "idea" for a solution to a problem with which we are faced: an "idea 2." In this second sense, we shall consider what you can do about an idea. In the first instance, you may get an idea 1 for a new and wonderful "GADGET" that meets a universal need. The safety pin was such an idea 1. So was the zipper.

Now the zipper was a wonderful—and original—new idea for fastening things together, but the problem was how to make it work—and be cheap enough to sell. To mesh properly and release easily, the little teeth of the zipper must have carefully designed shapes. Look at a zipper closely. You can see that the teeth fit together closely. To get such a good fit, they must be made to quite precise dimensions. And the teeth themselves are quite small! Furthermore, the slide must work smoothly. It is a marvel of simplicity and ingenuity. Yet, despite the necessity of carefully designed and precision-made parts, costs must be held down. Why? So that the market for zippers can be very large. To get low costs, raw materials must be chosen so that the precision requirements can be met initially and over the life of the zipper. The materials must be processed, and the hundreds of tiny parts must be assembled. To do the processing and assembly, completely new machines had to be conceived (idea 1's) and designed (a sequence of idea 2's). In other words, given the original idea of a zipper-type fastener, design problems had to be solved. To solve them, many "idea 2's" were necessary.

As you can see, the two concepts of an idea 1 and an idea 2 are quite different. And, naturally, because of the difference you must use different approaches to get the two kinds of "ideas." We shall distinguish carefully between the two kinds throughout this book.

Now what do we mean by the word "thinking?" Actually, in thinking you replace one mental picture by another mental picture. Thus, in thinking, you use your mental abilities. More particularly, you use your brain for a purpose—not for daydreaming.

A few of the important results of thinking are the following:

1. Getting an idea for something new or different.

2. Solving a problem and getting an answer in a satisfactory or useful form.

3. Making a decision by choosing between two or more alternatives.

As you will see, there are other applications of thinking. But these three are essential in the solution of any design problem.

The stuff of which ideas are made and thought about is "information." *We think about what we know or think we know.* And what we know or think we know we will call "information." A more precise definition will be given later and discussed at some length.

What do we do when we think? We select items of information stored in our brains; arrange them in patterns; combine them; rearrange them; discard some items; remember some new items; revise the patterns—in fact, we "process" information. You write a letter; you multiply 1246 by 761; you edit and correct your child's lesson; you struggle through your income tax report. You process information in doing all of these things. And many, many more. Again, more meaningful definitions will come later.

Our brains, and only our brains, can process information. We do not think with objects or energy, as such. Perhaps we can help ourselves by pencil and paper or by a device such as an abacus or a slide rule. Sometimes an adding machine or even an electronic computer may be useful. These are only tools to aid our information processing: our thinking.

There is no foolproof mechanical procedure for intelligent thinking. However, if you know an orderly set of steps to guide your progress in thinking, then you can be sure that the following will apply:

1. You will know where you are in the process so you can tell what information you need at that point.

2. You will have a framework to guide you in acquiring and processing the information you do need.

3. You will have an indication of the kinds of information you need for each step. This will help you to make checklists. And checklists can improve your performance.

A shot gun approach or a try-this-and-try-that scurrying around can be sheer waste of time and money. There are right ways to tackle problems. There are many wrong ways.

You need to know how to make a logical choice of the techniques for each kind of step in the solution of the various sorts of problems.

The penalties for wrong information, wrong thinking, or a poor decision can cost time or money or both. And time is money. In small matters, small amounts of money; in large matters, many millions of dollars. Poor thinking and poor planning have cost our country hundreds of millions of dollars on projects abandoned before they were ever completed. When you read your daily newspaper you are constantly reminded of the unhappy facts. A new missile system is designed, a few are installed. Then the whole picture changes. Much of the vast effort was in vain. Hundred-million-dollar battleships are built, then mothballed, and finally renovated for tens of millions more to go into limited service. New types of airplanes are proposed, worked on, and finally abandoned. And so on.

An orderly procedure can help you use your experience so you can handle new situations where you have had little or no background. After all, college doesn't teach all that you need in your career. You may shift or be shifted from one problem area to another overnight. And after the shift you will be expected to produce quickly. How can you protect yourself against such contingencies? Only by knowing the fundamental procedures for handling ideas.

Furthermore, you can test your thinking and be sure you are right before you act. A hasty decision can be dangerous. So can a wrong one. In fact, at times, there may be little difference between them.

Finally, if you are a manager, you can check on, control, and advise your subordinates better if you are orderly in your own thinking.

Once you understand the concepts and use the procedure in your work, you can expect a much better performance. You will be more efficient and accurate.

When you think about the problem, ideas and thinking are the common denominators of all that is new—of all progress. This is true not only for the great breakthroughs of science, but also of our day-to-day work. Everyone needs ideas. Everyone thinks.

Progress in *every* art and science depends on new ideas and careful thinking about them. Our physical scientists and engineers (and particularly our design engineers) have been responsible for changing our way of life from that of the 19th century.

Don't forget our other scientists: social scientists, psychologists, economists, and biologists. In each of these disciplines, men with new ideas and new thinking have made enormous advances. Freud and Keynes are just two of the great leaders. And don't forget government either. New forms are being conceived and thought about. The number of monarchies is declining. Other forms are being tried. Some of the newer forms are still being revised as the result of experience. Today, communism is different in many details from the teachings of Karl Marx.

In still another field—that of business—new ideas and new thinking are also apparent. Literally tens of thousands of quick-service food centers have been franchised in the United States. Supermarkets get larger and larger—yet modest-sized chain food stores with one or two clerks fill a need for late-hour items. At the other extreme, large corporations become larger by taking over noncompetitive enterprises and turning themselves into "conglomerates."

Thus, no matter what the field, the key concepts are "new ideas" and "thinking." Without them, progress is impossible. We all need to know the techniques of getting good "new ideas" and the best procedures for thinking them through. Whether your main interests lie in art, science, government, or management, you need to know and understand them.

The concepts are presented with little mathematics. Why? Because at the present state of knowledge, they can only be described qualitatively. True, block diagrams can be drawn to represent the conversion of input information into output information. These diagrams are helpful and enlightening; hence they are included. But when you try to describe the conversion operation in rigorous mathematical terms, the going gets sticky. After all, the human brain still is a mysterious thing. We don't know all about how it works; in fact, far from all! Getting ideas and thinking them through can be described in operational terms—but not in generally useful mathematical forms. The final chapter will make this point clearer to you.

So what do we do in our explanations? Some gifted mathematicians have proved some basic theorems about "information." Their theorems can be useful in explaining the procedure of "information processing." They can provide helpful guide lines, but that is all. And that is the way they are used in this book. The explanations are qualitative—not put in the form of complicated equations: they reflect today's "state of the art."

Why Study System Design?

A designer creates something that has never been. So does an artist, a sculptor, a composer of a symphony, or an author of a book or play.

And given his idea 1 of what is wanted, each of them must go through the same steps in creating his end product. For this reason, the techniques to be discussed are applicable in the arts and letters, as well as in technology. They can be useful in attacking complex problems in the social and behavioral sciences as well as in engineering.

Unfortunately, some of the key words used in this book have been applied to convey quite different concepts. As an example, consider the word "system." You can buy a book that describes a system for winning at blackjack in Las Vegas. A space rocket is also called "a system" or a "combination of subsystems." Hence, to avoid misunderstandings we need some definitions.

First, what do we mean by the word "system" in this book?

A *system* is a set of components arranged to perform some wanted operation(s). In the systems that we shall discuss, the wanted operation may be (1) a change or changes of a physical object (or objects), or (2) a change or changes of an item (or items) of information.

The word *wanted* is significant because it indicates that someone must conceive the wanted operation. They must have an "idea 1." Thus, conceiving the want is always a first step. Someone must want the results. For our present purpose, a design that nobody wants is just a waste of time, effort, and money.

Fig. 1-1 shows the simplest possible system. By definition, this simplest system has only one input and one output. The system itself performs only one wanted operation on the input to change it into the output. You push the button and the light lights.

A study of the properties of the simplest system can teach many important facts. For example, you can find out a great deal about the components that must be present inside the block labeled "system." These will be discussed later.

Actually, there are few, if any, systems that can be represented by Fig. 1-1. In real life, systems always have more than one input and more than one output. Furthermore, they perform more than one operation. Hence, a more realistic block diagram to represent a system is shown in Fig. 1-2. You can see that such a system has many inputs and also many outputs. The designer of a complex system is always studying one or more of the astronomical number of combinations of

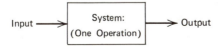

Figure 1-1.

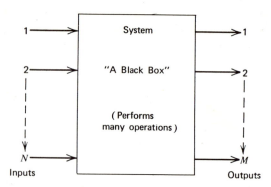

Figure 1-2.

components that may be so simply represented by the box labeled "system" in this figure. In the concluding chapter, we discuss the human brain from this point of view.

Don't be misled by the definition of the word "system." By intent it covers *any* technological system no matter how simple or complex. For instance, a motor, a lawn mower, an automobile, a house, an aircraft, a radar, a rocket, and a missile.

But it covers much more. Take a book as an example. The author first gets an idea. He then chooses and puts together the words and sentences to convey his message. Thus, he processes information. If he is wise, he tailors his message to an intended audience. Only if he does so will the audience *want* his work enough to pay to read it. Thus, there is an input, a processing, and a wanted output: thus, a system operation.

Or take a painting, a ballet, or a play. Again, the artist, the choreographer, or the playwright has an idea. He processes his information. And if he is successful, the output is wanted and applauded by the audiences who view his work.

Similarly, an outstanding opera or symphony is composed, presented, accepted, and acclaimed.

Furthermore, a new technological system may be worked on for months and years, and then but few people want the result. The Edsel automobile is only one example. U.S. military hardware furnishes many others. Likewise, a book may appeal to few readers; an artistic creation, to few viewers; a painting, to no buyer.

To repeat, by intent, the definition of the word "system" applies to the basic kinds of operations that result in many forms of innovative work.

The dictionary defines *design:* to map out in the mind; to plan mentally; to conceive of as a whole, completely or in an outline; to organize a scheme for doing something. Further: a design may arise in the mind for something to be done or produced; or for a mental project or scheme in which means to an end are laid down. Thus *design* refers to adaptation of means to an end; to the coordination of separate parts or acts to produce a result.

Notice the words "mental" and "mentally" in the definition. Design is the work of a human mind or minds.

You have a *problem* when someone wants a certain state of affairs and does not immediately know how to get it. Some initial knowledge is always available. But with a real problem the knowledge of how to proceed is imperfect (at best) or incomplete. There may be insufficient technical knowledge or inability to visualize the relationship of all design variables. Or just not enough information.

In some of the literature, a *problem* is defined: an unwanted effect; a deviation; something to be corrected or removed. If a deviation concerns nobody then it is no problem. This definition is often used by the top-level managers of a systems project rather than by the designers. However, a designer may have a problem that meets this definition. Such a problem cannot be solved unless a cause for the deviation is known or found. The cause may turn out to be a single event or a combination of events. For example, when a faucet drips, you suspect a washer is worn out. You check to see if you are right.

In solving such a problem, you use what knowledge you have to start

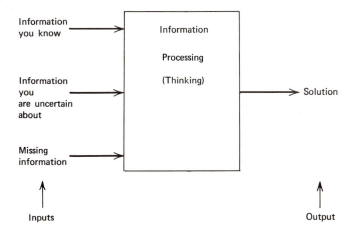

Figure 1-3.

an exploration. The exploration discloses more knowledge that causes more exploration. And so on until a solution is discovered. This type of problem solving is aimed at finding causes.

Problem solving can be expressed in the form of a simple block diagram (Fig. 1-3). As you can see, initially there are three kinds of inputs. Some information you have. Other information, you have but are uncertain about. Finally, you lack some necessary information.

Clearly, before you can really go to work, you must be sure that the uncertain information is correct. Furthermore, you must supply any necessary missing information.

After you get sufficient information, then you start processing it in a meaningful way. You go over the information; you organize it and reorganize it in different ways. Finally, you come up with the wanted output; an idea or solution to the problem.

In Fig. 1-3, the information processing is another word for what is often called "thinking." The mental picture that is changed contains both what you think you know and what you really know. In your thinking you reorganize the existing knowledge and get a new mental picture of the area in which you are interested. This new picture may contain more or less information than you started with. Quite often, the amount of information is the same; only the organization is changed.

Some people call such a reorganization of existing information "creation" or "creativity." Other people assign quite different meanings to these two words. The important thing to remember is that the thinking about information always involves the processing of information. Just that, and nothing more. What you are trying to do by thinking is to improve the organization of your initial information to come out with a new idea or a solution to a problem.

In the literature and particularly in psychology, you find definitions of two kinds of "thinking." The first kind is sometimes called "classical problem solving" or "convergent thinking": you end up with one answer. In fact, you have to come up with one answer *and no more*. Furthermore, your answer is either correct or incorrect. You have to discover a result that will satisfy explicit and immediately apparent standards.

An example of such a problem is the game of anagrams. Rearrange the letters:

TEREBLAY

so that they form a common word. When you get the answer you will know you are correct. But Queen Victoria is said to have spent several restless nights trying to puzzle out the right word. This is perfectly understandable: the eight letters can be arranged in an enormous number of different ways. Yet there is only one right answer.

The situation is quite different in "divergent thinking." In divergent thinking, you are asked to generate a lot of ideas—not to solve a problem by finding one final solution. Actually, there are no correct or incorrect "solutions." You have to take what information you have (complete, incomplete, or even wrong) and then choose from this information some ideas that meet some vague and often unspoken standards. In fact, you are not really solving a problem at all. You are using your brain to generate a lot of alternate answers. For example, what steps should be taken to rebuild our central cities?

In the world of advertising, some authors have called the divergent thinking "creative" or "brainstorming." In advertising, this type of thinking has both strong advocates and equally strong detractors. Some detractors have called it "cerebral popcorn." Critics of "creative" problem solving point out that it begs the question: How can you possibly select the best and most rational action?

Convergent and divergent thinking are necessary at different points in the step-by-step process of designing a new system. Each has its place and must be used at the right time. On the other hand, if you do divergent thinking when you should be doing convergent thinking, you are in trouble. Real trouble. And vice versa.

The very complexity of some of our present and future technologically based systems makes it necessary to understand more about the system design process. The record shows clearly that costly mistakes can be made. The cost of mistakes demands that the steps and procedures of the design process be closely examined.

Insight into the important factors can better be gained by study of the process of technical design than by experiment. For one thing, there are many necessary steps in getting out a complex system design. Furthermore, for some of the steps, many people may be involved—even tens of thousands. And finally, the development itself may take years—perhaps five or ten years. These basic facts make it difficult to design and carry out completely meaningful experiments.

Some of our large research and development laboratories cover the entire range of activities from the conception of the initial idea to the construction of the first model. However, the people carrying out the various design steps do not carry meaningful titles. In other words, the exact nature of their work is not necessarily obvious from an organization chart. Nevertheless, by study and analysis, the steps in the design procedure and their order can be identified. This is the method we have followed.

You should understand that there are considerable differences between system design as it is practiced today (and must be practiced) and the training given to students in our universities—even to graduate students. The great inventor, Charles Kettering, said essentially that if you are a student and flunk once you are out. There is no second trial. But an innovator or a designer almost always tries and fails at first. Then he revises his approach and tries all over again. The two situations (study and examination, and invent, try, and revise) are quite different. Our job in the universities is to teach students how to fail intelligently and then to keep on trying and revising until success comes at last. Progress in design depends on revising, revising, revising!

In practice, a complex system design is a long-term endeavor. This fact makes it easier to identify the steps that must be taken to achieve success. In school, we can experiment on one or a few human beings for one or a few class periods. This puts many difficulties in the path of seeing exactly what is necessary to succeed. If you substitute animals for humans, any tentative conclusions may or may not be valid in the real world. But we are not decrying what has been discovered by psychologists and others in their carefully planned experiments. We are merely stating that careful examination of a complete information-handling sequence can throw additional light on "ideas" and "thinking."

By studying the system design procedure, we can determine the kinds of information needed for each step. Usually the necessary information must be obtained from some source. How? You will see what is really involved.

The information must be worked over and processed by using logic or computations. The results must be evaluated, and decisions must be made.

For each step in the design process, particular techniques are most appropriate. These will be explained.

To show you how the step-by-step procedure works, the last chapter of the book applies this procedure to set up a model of the human brain. This worked-out example of the procedure also shows areas in which our present knowledge is inadequate. Hence the design cannot be completed—which often happens!

In this book, we shall concentrate on the work that goes on during the design period. We shall not consider the problems of producing or operating a system, or maintaining it after it is built. We are not minimizing the difficult problems that confront the people who produce, operate, and maintain systems. In fact, they use many of the same techniques as those who do the design. But their purposes are different—not their methods of thinking. Therefore, much of what is said about design-type thinking can be carried over to their work and also to planning new operations.

Given—Find

Take a far-out new system as an example of a design problem. A voyage of men to the moon will serve quite well. A few years ago, such a system was only a subject for science fiction. Nevertheless, we confidently expect that men actually will go to Mars and return.

Jules Verne described a moon trip in detail more than a century ago. But the necessary knowledge and ability to get there and back had to be built up step by step. Only a great nation (or an extremely large enterprise) can even contemplate creating such a far-out system.

In the last few decades, a number of such systems have been developed and made to work. Moon rocket systems are only one example. The point is that the steps necessary for such an achievement can be found by study and analysis of how these systems were designed and built.

In system work, problems almost always can be put into one of four categories:

AN EXISTING SYSTEM	A NEW SYSTEM
Some improvement or betterment	Some new combination of objects, information, or energy that either exists or can be made to exist
Isolation and remedying of some trouble in a working system	Some new use of an old combination

For an existing system, the betterment might be the reduction in size, weight, or cost. Improvements in performance or appearance also fall into this category.

When an existing system gets into trouble it is necessary to determine what is wrong and then to fix it. You turn your automobile over to your repair shop to do exactly this set of operations.

The situation is quite different when you want to design a completely new system. Very seldom do you start from scratch. For example, you may have to use available materials: iron, steel, brass, and copper. Also you may have to develop new materials to reach your objective. Ultrahigh-speed airplanes have their wings and fuselage made of stainless steel or titanium. But these are unusual cases. For many—perhaps for most—new systems you merely select and assemble a new collection of existing parts to perform the new operation.

A system designer occasionally (but not often) must think up a new use for an existing system. And there may be hazards in even trying to do this. In your home workshop, you may substitute one tool for another when you do not have what you really need. But you know that the results may not be wholly successful. This also applies for larger systems. Occasionally, however, a new use can be found. In fact, the patent office recognizes such a finding as a legitimate subject matter for a patent.

Important types of design problems

In Fig. 1-4 an input object is changed into an output object by the system operation. To control the operation (if only to show when it is wanted), an information source is needed. This figure represents a basic type of system operation.

In Fig. 1-5, input energy is changed into output energy by the operation of the system. Just as in Fig. 1-4, an information source is always necessary. This is a second basic type of system operation.

Also, in Fig. 1-4, the input energy may be carrying information that is changed in some way by the system operation. Again, an information source is necessary. This is the third basic type.

Thus the three basic types of system operations change are as follows:

1. Input object(s) into output object(s).
2. Input energies into output energies.
3. Input information into output information.

For a particular complex system, some combination of the three basic types may be necessary.

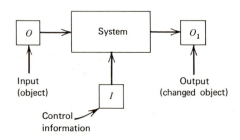

Figure 1-4.

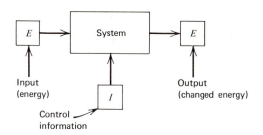

Figure 1-5.

In many complex systems, operations on both input objects and input energy are performed. Thus, these systems are made up of combinations of the basic types.

These simple diagrams permit some important observations.

In Fig. 1-4, all of the input mass (wanted or unwanted) must appear at the output. In other words, mass must be conserved. An exception to the rule occurs when mass is changed into energy as in the hydrogen bomb. Many manufacturing processes generate unwanted scrap material. Planning for the utilization or removal of these by-products may be an important part of the design problem.

Similarly, in Fig. 1-5, input energy must equal output energy, according to the law requiring conservation of energy. This law holds true whether the input and output energy forms are the same or different. In other words, it holds true even if an energy conversion (for instance, from mechanical to electrical) is performed by the system. You run into the energy balance concept in the design of steam power plants and of electrical motors and generators. All of the input energy must be accounted for as outputs.

A usual form of unwanted output energy produced in all kinds of systems is heat. In the early models of high-speed computers, removal of the unwanted heat was a major problem. Large amounts of refrigeration were needed.

In Fig. 1-5, if the input energy is carrying information to be delivered at the output, Shannon proved that the total output information can never exceed that at *all* the inputs. In many practical cases, it is less; in some systems, it is very much less.

Table 1-1 is based on the simple diagrams of Figs. 1-3 and 1-4. It shows that for either figure there may be *eight* different kinds of design problems. Not one—but eight!

The problems differ in what is given and what is to be found. In the table, only the three factors of output, input, and operation are entered. The information required for the operation and indicated in the figures by the block labeled I is omitted from consideration, although it always must be carefully specified.

Consider the change of objects first. This may be called the *manufacturing* problem. An input object is to be machined, heated, transported, or otherwise changed by the system operation. We are only told what the wanted output is to be. Except for the condition of conservation of mass, the input and the system operation are unspecified. In the design, you must choose a suitable input material and methods for changing it into the desired output.

Table 1-1.

CASE	OUTPUT	INPUT	SYSTEM
1	G(iven)	F(ind)	F
2	F	G	F
3	F	F	G
4	G	G	F
5	G	F	G
6	F	G	G
7	G	G	G
8	F	F	F

In Case 2, you are given an input and are to find a useful output and also some way of making the output: a *utilization* problem. For example, suppose your company has some by-product (perhaps a chemical or scrap material). You are told: (1) to find some use for the by-product; (2) to process it so that it will be useful; or (3) both.

In Case 3 (Table 1-1), you have a system. Neither input nor output is specified. In a factory, the system might be a lathe which requires certain kinds of input metals and a wanted output object that can be made from these metals. In terms of human beings, you have an engineer to whom you must assign work that he can do—work that will produce ideas or designs that you can use.

Cases 4, 5, and 6 (Table 1-1) are similar to Cases 1, 2, and 3—with one important difference. Now *two* facts are given and only *one* must be found. At first, you might think that the more information you are given the easier the problem will be to solve. Sometimes, yes. Sometimes, no. Given both the input and the output *and the wrong tools* you may not be able to make the output from the input. For example, a file and a wrench are not always interchangeable. If you have a file and need a wrench you may be in deep trouble.

You can easily find equally good examples for the other cases.

Now let us examine Cases 7 and 8. At first glance, these appear to be trivial.

In Case 7, everything is given; nothing is to be found. Thus you have no problem. Quite true. But let us go back to Case 1. Here you are given the output and are to find a suitable input and a suitable system. Your system design problem is to change, step by step, the situation from Case 1 to Case 7. In a later chapter, we shall show the few ways by which you can logically make this change.

Now examine Case 8. Here nothing is given. Everything is to be found. Again this case might be dismissed as trivial. But you can also look upon it as describing the situation when you need an idea. And all really new systems start with an idea. In other words, you must convert Case 8 into Case 1, 2, or 3. The idea to make this conversion possible is missing, and you must look for it.

The Steps in System Design

This book emphasizes the importance of both ideas and of thinking in design. Design always involves the following information:

1. Information in the form of facts and data.

2. Information in the form of patterns and relations between the items of *1*.

3. Information processing using *1* and *2*.

System design involves objects only if and when a model is finally built. Of course, you may use such aides to thinking as pencil and paper, tabulating machines and, possibly, a computer. Even more important, the purpose of building a model is to gain more information: to check whether the previous thinking has been correct and complete. An airplane is designed on paper and computers. A prototype model is then built and test-flown.

In any study of system design, the first question to be answered is: How does a new design start?

At the beginning of a project there are two possible situations:

1. An idea for a project does not exist. In this case, the first thing to do is to get an idea—to find a good problem and set up goals.

2. A need exists. You must find the best way to meet it. To do this, you must solve a set of problems and get satisfactory answers.

In *1*, you must find out what you want to do. In *2*, you must find out the best way to do it. Only after you find out what you want to do, can you start to find out the best way to do it.

Thus, the sequence is *1* and then *2*; or simply *2*.

In proceeding from what you are given (*1* or *2*) to finding your solution, you want to take all the necessary steps so that your work is complete. *No omissions.* But you want to take no more steps than are necessary, because this would mean an inefficient procedure.

Hence, to get from "given" to "find" you need a well-thought-out program of ordered steps that may be called a "design procedure."

Later, these steps will be identified, together with the reasons for the ordering. As you read further, you will find each step discussed at length.

True, as the design work proceeds you often have to return to an earlier step in the sequence and make revisions: you have to "recycle." Such recycling is a characteristic of design. But, eventually, each step in the procedure must be taken in an orderly manner—perhaps repeated several times. Recycling and revision does not mean that any step in the sequence may be omitted. Quite the contrary! Steps are repeated—not omitted! And the steps cannot be executed in any random order. First things always come first! Last things, last!

The idea

The first step in any design is getting an idea 1. The idea 1 may call for something new, a new use for something, or a change in something that already exists. Often (but not always) change is intended to be a betterment. Sometimes it is merely "a new look"—a change in appearance. In the automobile industry, a new look without other major changes is often called "face-lifting." Sometimes a new design is not intended to be better—only cheaper. A few years ago, Chevrolet brought out Corvair; Buick, the Special; and Oldsmobile, the F-85. These compact cars were to compete with European imports. All sold below the regular price ranges of the sponsoring General Motors Division. You are familiar, from experience, with newer and cheaper models of home appliances, TV's, and radios. After all, not everybody can afford a Rolls Royce.

As another example, consider the laser. Now the ideas that led to the creation of the laser were truly brilliant. They resulted in the ability to generate coherent light energy for the first time. Tens of millions of dollars of research and development money have been devoted to further exploration. But applications have been few and far between. Some unkind (and inaccurate) cynics have quipped: the laser is an answer in search of a problem. Actually, the laser is in search of an idea 1: a *need*. As time goes by, perhaps ideas 1 will be found in large numbers. And possibly not. In any case, the costs have been incurred.

After an initial idea has been proposed, often studies are made to generate other new ideas. The original idea is worked over, modified, and worked over again.

Then several ideas are available for consideration. Before going further, it is necessary to choose the best proposal. Then the objectives can be set up for the design.

Rules of the game: constraints

When you play a game, you must play according to the "rules." In football, you cannot make a touchdown by running the ball down the sidelines out of bounds. In baseball, local ground rules may say that "over the fence is out," or they may say that "over the fence is a home run." Thus the rules may depend on agreed-upon circumstances.

A somewhat similar situation may exist for a system design. Sometimes the rules are not called "rules." They may be called "objectives and goals," "tolerances," and "the environment." But no matter what they are called, they are really the rules of the game. Only after objectives and guidelines have been established can design work be started with confidence.

Objectives and Goals. Objectives and goals guide and limit your design work. They may be of many kinds. They can be frustrating. But nevertheless they are there.

Take a wonderful new idea as an example. At least, an idea that seems wonderful to you. Top management either consciously or unconsciously thinks about it in terms of the company's policies. They ask themselves questions such as: Does it fit in with our plans? Does it mean a new line of products? Does it fit in with our image of producing top-quality products or our image of "we will not be undersold?" Do we have enough resources so that we can go ahead with it? Are they the right kind of resources? Will it tie up capital we need for something . else? Is enough design talent available?

When faced with a new idea, some managers almost automatically react with a stand-pat attitude. After all, it is much easier to say

"This just is not the right time" than to sit down and take a hard look. Therefore, do not be discouraged if your ideas get little attention.

Objectives and goals can be classified as follows.

1. MUSTS. The set of performance and other requirements that *must* be met. For example, you may be told that your design must use electrical energy. Hydraulic and mechanical devices are not acceptable.

2. MUST NOTS. The set of constraints stating what the system *must not* be or do. For example, you may be told that if your system costs over $10 to make there will be no market.

3. WANTS. These are not hard and fast requirements; merely
4. DON'T WANTS. what the words say.

In the course of analysis of a proposed design, you may find that you can make the system a bit cheaper if you add a little weight. In either case, you meet all the MUST and MUST NOT requirements. But one alternative may be preferable over another because of the WANT and DON'T WANT constraints. The operation of increasing weight to get lower cost is often called a "trade-off." The study of possible trade-offs is an important part of the design procedure.

In most situations, a MUST is really a MUST; a MUST NOT is really a MUST NOT. In most situations, but not in all. Why? Because breakthrough design may follow an insight that discloses that a MUST or MUST NOT either is not so or can be gotten around. But successful breakthrough design is the exception—not the rule. The financial losses can be staggering when the breakthrough is illusory. The first unsuccessful attempts to build variable-swept wing aircraft provide an example of the hazards.

Tolerances. We have already seen that a system has a set of inputs and a set of outputs. The system performs operations on the inputs to change them into the outputs. Thus the inputs, outputs, and system performance are intimately related.

In any practical system, the inputs will show variations. These may be variations in physical properties or in energy. The variations may be small or they may be large.

Similarly, in the operation of any practical system, the processing of the inputs will also vary. Tools wear out as they are used. Consequently, the parts made by them change. For a particular system, the acceptable variations may be quite wide. For example, they may be expressed in inches or feet. For other systems, they may be extremely small. For example, they may be expressed in thousandths or even much smaller fractions of an inch. Optical systems, such as microscopes and telescopes, require extreme precision in their manufacture and assembly.

Depending on the system (and, in some cases, even on the components of a system), the tolerance may be expressed merely as being a quantity less than some value, or more than some value. In other cases, the requirement may state that the quantity be within certain limits, say within 1% of a particular value. For example, you expect the voltage at your electrical appliances to be within a few volts plus or minus of 120 volts, and the frequency close enough to 60 cycles so that your electrical clock keeps acceptably good time.

To meet the wanted output tolerances, you, as a designer, must know what can be accepted. Then, you must allot the acceptable variations to the system and to the system inputs. Either the system performance or the inputs may be allotted a larger share of the overall tolerance. Whether one or the other is favored often depends on the relative costs of improvement of the system or controlling the inputs more closely. For example, you may have to take your inputs as they come to you. You may have little or no control over them. This can be true if the inputs are (for example) products of a mine, forest, or farm. In such cases, you cannot control your raw material input as closely as you would like. The system must be designed to take care of the variations. Tolerances can be classified in the same way as constraints:

1. MUSTS.
2. MUST NOTS.
3. WANTS.
4. DON'T WANTS.

Also, the WANT and DON'T WANT tolerances often permit design trade-offs.

The Environment. Every system, every designer, and every decision maker operates in an environment.

By definition, the *environment* consists of everything outside the system or human being that either affects the performance or is affected by the operation, or both.

For a physical system, you need to consider only the physical factors of the environment. For example, these may be temperature, atmospheric pressure, and humidity. For a human being, a designer, or a decision maker, both the physical and mental environments must be considered. In some ways, people are more tolerant of environment than machines. But in others, they are far less tolerant. A machine can work and disregard a hostile emotional environment. A human being may be completely frustrated and able to do nothing.

As a designer, you must carefully consider all of the effects of environment upon your design. If the environment varies with time, you must take this into account. After all, your automobile is expected to work in the ice and snow of winter as well as in the heat and rain of summer. It must operate in the dust and smoke of the large city, and also in the dust storms of the desert. If it does not, then it is a poor design. In other words, it must work acceptably well over the full range of all of the variations of its environment.

In a spacecraft, the men must be provided an environment in which they can live and do their tasks. Outside is a nearly absolute vacuum and subzero temperatures that never occur on earth.

Another point: the environment during transportation and storage may be more severe than those encountered in use. A spacecraft being carried to the launch pad may be stressed more than it will be in outer space. A truck destined for our armed forces may be subjected to the heat and humidity of the tropics or the intense cold of the arctic. It may be sprayed with salt water during its overseas trip. The designer must consider all of these possibilities.

Fig. 1-6 shows a simple block diagram of a system and its environment. It indicates the inputs from and the outputs to the environment.

For a sizable system, the conceivable number of inputs and outputs may be extremely large. In practice, as a designer you must select those that you believe to be most important. In doing so, you divide the possible interactions into those which you know are important and those you believe you can safely neglect.

Different designers may choose differently. You must use your own judgment. A correct choice is extremely important. If you neglect an important input or output, the result may be disastrous. Airplanes have crashed because someone neglected (or did not investigate fully)

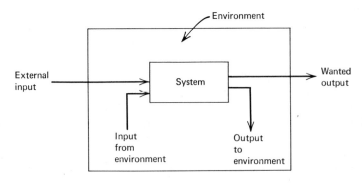

Figure 1-6.

the effect of a particular atmospheric condition on a wing. The wing vibrated, fell off, and the plane went down. Bridges have collapsed because wind effects were not completely analyzed. There are perils in taking too many effects into account. Why? Such caution means extra work of no value, except possibly as insurance. And, as you know, insurance always costs money. Insurance in design is no exception.

To sum up, if you neglect an important environmental condition in your thinking, you may end up with disaster. On the other hand, if you take too many conditions into account, you may spend too much time and money. Then, top management will let you know in no uncertain terms.

It cannot be overemphasized that proper evaluation of the effect of a system's environment is a very important part of the design and management functions.

Synthesis

Only *after* the objectives and goals, tolerances, and environmental requirements have been established, can the design work really be started.

The first step in design is *synthesis*.

What is *synthesis?*

Synthesis involves finding *any* collection of objects that can perform all of the wanted functions. Futhermore, the collection must meet *all* of the requirements of the specifications.

What must you do to find and put together such a collection?

You must gather together all of the available information about the objects you intend to—or would like to—use. You must organize this information. You must classify it. You must be sure that objects you want to use will work together.

If you are wise and able to do so, you will come up with several tentative collections that promise to do the job.

Analysis

What is *analysis?*

In system design, by analysis we mean a very particular operation. You must know and understand it well. By analysis, we mean that any tentative collection must be studied and checked to make *sure* that it meets *all* of the objectives.

More often than not (based on years of experience in checking system proposals), the first synthesis will be found to have weaknesses. Sometimes the weaknesses are so great that the system is useless.

Almost always, one or more tentative collections are chosen as possible final choices. At this stage, you are trying to get an optimum or best all-around design.

Selection

After analysis comes *selection*. Studies have shown one (or hopefully more than one collection) that will meet the objectives. Now, the best alternative must be selected.

Sometimes, in large organizations with ample money and facilities, two alternatives will be authorized for further study. This is unusual. Ordinarily, only one can be pursued further.

How do you select the best idea? You go back to the rules that were given you along with the initial objectives. You compare each alternative to the objectives and use the rules to make a choice. Perhaps the rules say that the lowest cost collection that meets the requirements is to be chosen. They often do say so. But for one exception, in space exploration cost may be given lesser importance than space occupied, weight, or reliability.

Sometimes the choice between alternatives is clear-cut. Sometimes a choice must be made on the basis of minor differences. But sooner or later a choice must be made.

Decision and action

In problem solving, your aim is to find an idea for obtaining a solution. In decision making, you aim is to find the proper action to take. Actually, taking the action is a second step that will be discussed later.

Fig. 1-7 illustrates the decision-making operation. Just as in problem solving, all the inputs are information. This input information is reorganized to produce the wanted output: a decision.

As an example of a simple decision-making operation, take crossing

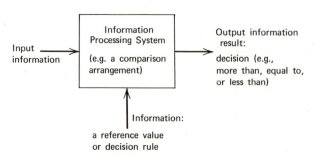

Figure 1-7.

a busy street at the corner. First, you check the traffic light. If it is green, and if a car is not about to make a right turn into your path, then you step off the curb and start your crossing.

Communicating the results

After the selection of the best alternative is made, what next?

You must tell other people about the results of your work. You may do this orally. But most of the time, you will have to prepare a convincing report. More about this later.

Remember, unless you are a one-man operation, you do not decide what to do next. A manager does. So you tell him what you think ought to be done.

A model?

Why build a model? Simply because all human beings are fallible. Electronics engineers build a "bread-board' model of a new circuit. Often they build a "brass-board" model with the exact layout and proportions of the proposed system. Why? Experience has taught them that even experienced people cannot foresee all of the possible contingencies. Mathematics cannot take into account all of the possible variables. Something may be overlooked in their thinking, something important—possibly even something disastrous.

If you are going to make thousands or millions of systems, an oversight cannot be tolerated. After all, would you be willing to fly an airplane that test pilots had never taken off the ground? Almost certainly not. Even after the models are built, even after the test flights, airplanes must be certified by a federal agency. People's lives are at stake. Some of the great mathematical brains of the world are used in the design of our planes. But mistakes can be made. People's lives cannot be jeopardized by mistakes.

Actually, information gained from a model almost always requires a rethinking of the entire design problem. You may have to come up with a different collection of elements to meet the requirements. You may have to restudy the original collection or the one that supersedes it. Sometimes—but not often—the whole set of objectives must be changed. Maybe they cannot be met; maybe they cost too much; maybe the system weighs too much; or maybe it just will not work.

Complex system problems

Often, trying to solve several different problems all at once is quite a hopeless undertaking. It is particularly difficult if the different problems require different input information and different processing. One author uses the word "mess" to describe such an unhappy attempt. Another, who apparently has had some experience, calls simultaneous problems "a can of worms."

Although various authors use different words to describe the multiple problem situation, all of them agree that to solve complex multiple problems you must break them apart: factor them. Once factored, you can attack each piece individually. Hopefully, the solution of all the pieces will give you the wanted solution. This idea occurs in the literature over and over again in many different contexts.

As an example of a fairly complex system, take a house. There is the structure itself, which must be suitable for the environment. And the environmental conditions not only include the immediate surroundings (and any zoning laws) but also such factors as the temperature range, water supply, electric power sources, and sewage facilities. Besides the structure, there are a number of subsystems: for plumbing, electricity, heating, air conditioning, and so on. Careful planning is necessary for a modern residence, so that all subsystems are properly designed and located. Changing one subsystem—say adding a room— means revisions of several subsystems: a recycling of the design.

The End Results of Design

The mental processes that we are calling system design result either in:

1. A new object.
2. A collection of objects, that is, a new system.
3. New information expressed in some form.

A successful design thus either produces objects or information. For example, the object may be a new chemical compound such as nylon. As another example, one producer advertises "Where new ideas take shape in aluminum."

As an example of a new system, take a new supersonic airplane or a rocket ship to fly to the moon.

Examples falling into the information category are a new musical composition, a new book, or a new mathematical proof.

Always, one result of the system design is a communication to other people. These may be the people who are to manufacture the object or assemble the system. Or they may be the people to whom the information is of value. They may be the audience at a symphony; the readers of a book. Thus the communication may be written or oral or both.

All system designs are not carried through to completion. In fact, only a small fraction are. Work may be stopped because it has become clear that the objectives of the project are not technically or economically feasible or both.

When a design is stopped before it is completed, it may disappoint those who have worked hard on it. Nevertheless they have some consolation. New knowledge has been gained.

Sometimes, the knowledge gained by failure may point the way to success. Or it may prevent further futile effort. In mathematics, a proof of impossibility is often a great achievement. For example, the proof that all angles cannot be trisected using only a straight edge and compass has saved much useless work. For centuries, well-intentioned people tried to build perpetual motion machines. When the law of conservation of energy was accepted, the hopelessness of their search was apparent. Today, even school boys know that perpetual motion is impossible.

Design Is an Unfolding

When you think about it, the design process is akin to the unfolding of a flower. It starts with an idea. The original idea is the bud. It may give no indication of the shape of the final system. As knowledge is acquired and organized, the future system gradually takes form. As more and more details are added, it is like the flower gradually coming into full bloom. And just as the flower is beautiful for a time and then fades away, the system is used and possibly admired for its life span. Then it, too, is retired and thrown away.

Thus the design of a new system is an evolution, a development. The results of the design may be revolutionary—or they may not. The design of the systems to convey men into space have taken years. Whether they will revolutionize the future of mankind on earth is not clear. We just do not know.

In your thinking, do not confuse design and discovery. To *discover* is to get the first sight of, or knowledge of. Columbus discovered America. Fleming discovered penicillin. You discover what has always existed but has not been known before. Furthermore, what you discover may be either good or bad. In contrast, you work, plan, and develop a design. And as we again point out, you design something that someone wants. It has not always existed and it is something that is wanted. Thus, in two ways a design differs from at least some discoveries.

A design may or may not involve invention. To *invent* is to find out, to combine by ingenuity, to originate. A second meaning is: to fabricate in the mind. Also, you invent new combinations or new arrangements not used before. Thus, many designs involve invention. But equally true is the fact that many designs do not involve invention.

The design of a new building or the plan for a new city is not a patentable invention. According to the legal definition, invention must be a step forward beyond the bounds of existing art. A design may be new in a practical sense and yet not new in the legal sense so that it is patentable.

Some inventions are the result of "flash of genius." A brilliant idea suddenly occurs to an inventor. It need not be worked out in complete detail before it becomes patentable. On the other hand, a design is as we have stated, an unfolding, an orderly progression from idea to final results.

Finally, both designs and inventions are products of the mind: of "thinking."

2

System Organizations

Functional Organization Charts

Graphical representations of a system can be (and often are) used to show how the system components are related to each other. Block diagrams are one form that experience shows are very useful for this purpose. Some examples were used in Chapter 1. Block diagrams are useful because they can give a great deal of information in a form that can easily be grasped. For this reason, they can be an excellent first step in a discussion of system design.

Blocks in such diagrams often—in fact, usually—are rectangular. However, other shapes such as ellipses and diamonds also occur. Furthermore, different colors, widths of borders, and shadings are used to help interpret complicated diagrams. Such differentiations are a form of "coding," which is defined in a later chapter.

Family tree diagrams

In the family tree or hierarchal type of organization (Fig. 2-1), the blocks are arranged in rows and subrows. Almost always, the most important block is at the top or at the left of the page. Whatever arrangement is chosen is decided when you make the chart. Such a chart is a logical method of representing a set of functions and some of the relations between them.

In many cases, the spatial arrangement of the blocks in a row has no particular significance.

Figs. 2-2 to 2-5 are typical examples of family tree arrangements. Fig. 2-2 shows a family tree. Of course, this is where the particular arrangement got its name. Fig. 2-3 shows a functional block diagram for a simple machine. Fig. 2-4 shows this type of presentation for a business organization; Fig. 2-5 for a university. You can see that, organizationally, there is a great similarity between a business and a university.

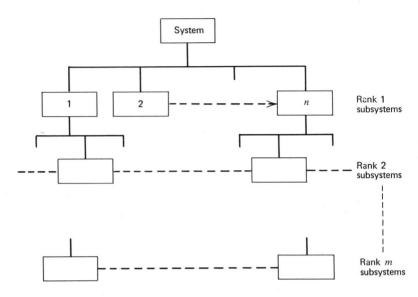

Figure 2-1

The family tree type of representation is very useful for showing a work-together type of organization. For example, in a business, the president, those reporting to him, and their people all work together to forward the aims of the organization. If any block is not filled, then that function is not performed. Presumably, part of the mission of the organization is not accomplished.

In some situations, certain blocks are active only part time. For example, in a very small business organization, a full-time auditor may not be justified. However, the auditing function should be shown on the organization chart. The wise top-level manager calls in an auditor to perform this function when he is needed.

Also, in a machine, you may find duplicate (redundant) parts or even subsystems to ensure reliability. Late-model cars have dual hydraulic braking systems. They also have a separate parking brake. Such redundancy is provided at extra expense to insure the reliable or safe operation of the machine (or both). Similarly, in business organizations, additional people may be put on the payroll to fill in if someone is absent.

Reverse tree organization

A reverse tree organization is also useful in certain circumstances. As shown in Fig. 2-6, initially there are a large number of competitors. In

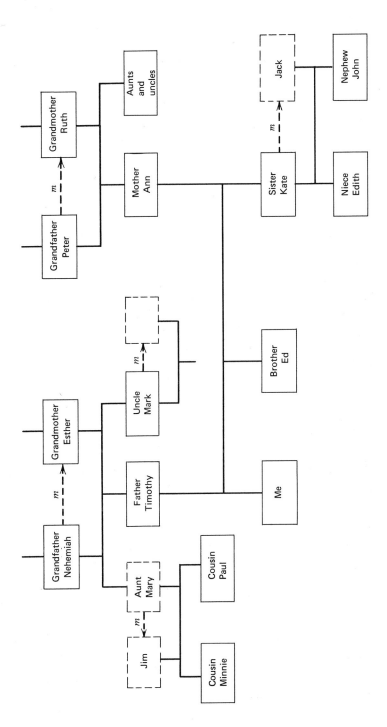

Figure 2-2.

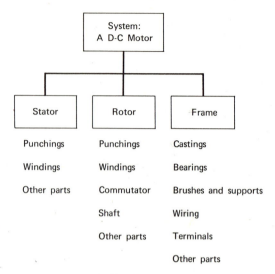

Figure 2-3.

some way, the choice is narrowed down, until only one remains. You have seen such a diagram made up for a golf or tennis tournament. The long list of initial contenders occupies the blocks on the left of the figure. As they are eliminated, their number decreases. Finally, one is the winner.

You will find that the same type of figure describes certain situations in system design. Initially there may be a large number of competing ideas. One by one, these are eliminated until the best one is chosen. In tennis or golf, one match eliminates only one player: the loser. In selecting the best alternative for a design, large numbers of ideas having some unwanted property may be eliminated at one time.

Ordering of functional blocks

As stated earlier, in some cases, the arrangement of functional blocks (or objects corresponding to blocks) is not important. In building a wall, the particular brick that you put in a particular place is not important. In this case, the unimportance arises because all the bricks are essentially alike. Consequently, the way that you arrange them is of no great consequence. The family tree type of representation imposes some constraints because all blocks are not essentially alike. In Fig. 2-4, the most important officer, the president, is unique, and occupies the top position in the chart. His vice-presidents occupy the next row.

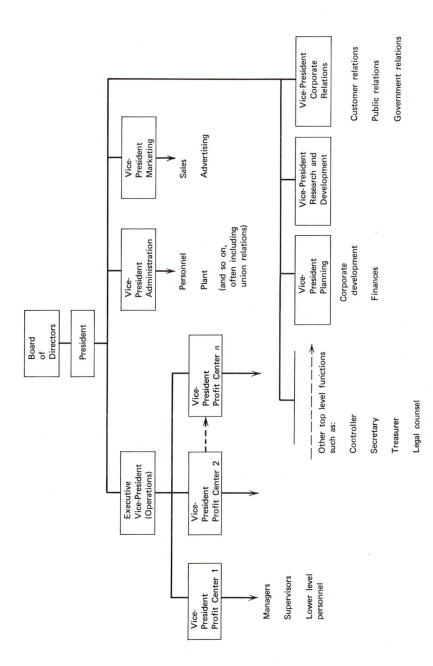

Figure 2-4.

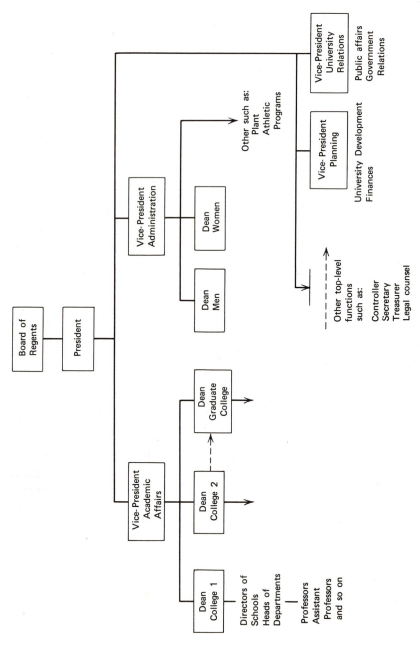

Figure 2-5.

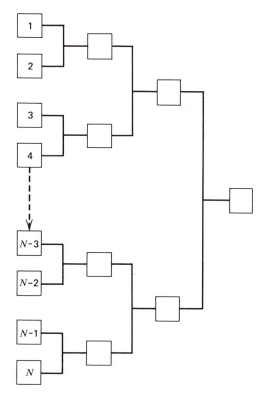

Figure 2-6.

Quite often, all are of equal rank. Position in the row has no meaning. But in some companies one vice-president is "more equal among equals." He may be put at the left side of the row or given a title such as "executive vice-president" to portray the actual situation more precisely.

In some functional diagrams, the placement of symbols may be very important. As a simple and familiar example, consider the diagram showing the location and type number of the tubes in your television set. Mix up the locations and the numbers and you can be almost sure that the set will not work. Also, in an assembly line, the right machines must be in the right places so that the operations can be performed in the proper order.

As another example, take the arrangement of the four letters r, e, a, and p on four blocks. You can shift the blocks around so that the letters appear in different orders and so that the letters spell different words. Fig. 2-7 shows anagrams of the four letters. Such one-after-

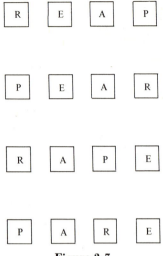

Figure 2-7.

another organizations of functional blocks may be called "tandem" arrangements.

An analogous situation exists in time. In some series of operations in time, the order is unimportant. Take a familiar example. When you add a column of figures, you can start at the top or at the bottom or some mixture of the two possibilities. If you want to make things hard for yourself, you can even start in the middle.

In other situations, the order of performing the operations is very important. Did you ever try putting a watch back together? Unless you have had training about the right way to do it, almost certainly you will have a few odd parts left over. In information-processing operations, order can also be important. When you prove a mathematical theorem, each step has its place in the chain of reasoning. Mix up the steps, and you have no proof.

A one-after-another organization is sometimes called "sequential."

Combinations of functional arrangements

Many complex systems use a combination of the various kinds of functional arrangements.

All of the parts must be provided for the system to operate as it is intended to operate. However, all component parts may not be in use all of the time.

An automobile engine provides a simple and familiar example. For a particular cylinder, intake, compression, expansion, and exhaust follow in a fixed order. The order is important. Perhaps there are eight cylinders in your car. The sequence of the four basic operations is fixed for each cylinder. Furthermore, each cylinder operates in its cycle with fixed relation to the other seven cylinders.

An automobile furnishes other examples of objects arranged to perform in fixed sequences. Take starting, for example. You don't take off on your trip with the emergency brake on—at least you shouldn't. Your automatic transmission is designed to shift gears for you under closely controlled conditions as your car accelerates.

You can find many analogies between starting, operating, and stopping your car and taking off, flying, and letting down in an airplane. However, the airplane has many, many more steps that must be performed in the right order. There are so many steps, in fact, that the pilot and copilot of a jet airliner have a list of the things to be done. They check each other as each step is taken. They don't depend upon their memories because an error might have dire consequences.

The designer's role

The requirements for the organization of the system functions (and hence of the functional blocks representing them) are studied and set by the system designer. This is part of his job.

Requirements for a Satisfactory Functional Organization

Functional charts are intended to be representations of logical arrangements.

In the early stages of a design, blocks may represent ideas of the wanted and necessary functions. In later stages, they may represent blocks of hardware intended to perform those functions.

Each kind of block diagram serves useful purposes as the work progresses. Quite possibly, a combination type diagram may be helpful. For some functions, hardware may be available; for others, not. A diagram may be drawn to show such a state of affairs.

Of course, logic (and particularly symbolic logic) is often thought of as a branch of mathematics.

In any case, to be a satisfactory representation, a proposed functional organization must meet three requirements. You may have come across two of the three requirements in mathematics.

The first requirement is *necessity*. Every functional block in your organization diagram must be necessary. If you use a functional organization diagram in your design work, you must be sure that this requirement is met. Otherwise, a block that is not necessary can be left out without affecting the operation. An unnecessary block almost always means increased costs, or increased complexity, or both.

Despite the general rule just stated, sometimes a functional block is left in. Possibly, it costs more to leave out the function it represents than to keep it in. For example, you may have to modify an existing equipment. It may have features you don't really need—but are already there. You may take them anyway. Otherwise, you would incur still higher costs by taking them out.

There is another exception to the general rule. You may add extra functional block—"redundant" elements—to perform a function: safeguarding the operation.

The second requirement is *sufficiency.*

The proposed organization must provide functional blocks for *every wanted* function. *None must be omitted.* You must check to make sure that this requirement is met. Otherwise, the organization is incomplete and cannot meet the system objectives. Finally, depending on the objectives, you may provide more than enough components. You may deliberately overdesign. For example, you must think about overloads or equipment failures. To make passing easier, your car has more cylinders and hence more horsepower than a bare minimum. The designer decides how much more to provide. Your car also has dual braking facilities so you can stop when you want to—or when you have to. Fuses are put in electrical circuits to prevent a disastrous result of a short circuit. In a factory assembly line, people are provided to take care of absences from their work stations.

As pointed out earlier, the *ordering* of the steps and blocks may be important. If so, as the designer, you have the responsibility of checking to make sure that such requirements are met.

Equivalent Organizations

A concept of equivalent organizations is important to a designer. Notice that we use the word "equivalent"—not "identical." We are using the word "identical" to mean that each and every functional block in two representations are exactly the same. Furthermore, if the ordering is important, then the ordering of the blocks is exactly the same. These are stringent requirements.

Two organizations are equivalent if they meet both of the following requirements:

1. Each must accept inputs 1, 2, . . . *m* which meet acceptable tolerances.

2. Each must deliver outputs 1, 2, . . . n which also meet acceptable tolerances.

Note that nothing is said about the internal organization of the system which performs the operations. The inputs in the two cases need not be exactly the same. They need merely to be within acceptable tolerances. Likewise, the outputs need not be exactly the same. But they, too, must be within acceptable tolerances.

The idea of equivalence permits a designer to come up with acceptable alternative proposals for his problem. After all, Ford builds an automobile that competes with Chevrolet. They each have engines, brakes, a chassis, etc. You may prefer a Ford, I may prefer a Chevrolet. We each get a system that furnishes us transportation. The differences between them is the stuff of which the advertisements are built. Most of the differences may be really minor. A telephone call offers another example. When you make a long distance call, the electric waves generated by your voice may be carried by a pair of wires, or in a coaxial cable or by microwaves. You don't know how they are carried—and you don't care. All that you ask is that you can talk to the person whom you called, and that you can understand his reply. To you, the transmission facilities are equivalent. You can think of many other examples from your own experiences.

Advantages and Disadvantages of Functional Charts

In the early stages of a design, functional block diagrams can certainly help you get your ideas together. They can be a graphic representation of your logical structure. Such a chart can be a tool and can help to put your vague ideas into a concrete form. It can be checked, it can be revised, it can be rechecked, and revised again.

For these important reasons, functional block diagrams are a valuable early step in a system design. Nevertheless they are only an early step. Why? Because a functional block diagram just cannot and does not give all of the information that you need to work out your design.

For one thing, such a chart does not show the inputs and outputs of each functional block. Knowledge of these inputs and outputs is essential for a complete understanding of the organization. For the kinds of functional charts we have just discussed, this may be difficult to do. For example, in a chart showing the functions of a business organization the lines that would show the communication paths can be numerous. If you try to show who talks to whom, the result can remind you of the mess an active kitten can make with a ball of yarn. You can find examples of such charts in the literature, particularly in the literature of management science. But the attempts to draw them have not been too successful. They just become too complicated to follow. In fact, for almost any complicated system design, another form or chart is highly desirable: a diagram showing the inputs and outputs of every block. Managers are beginning to understand the weaknesses of the usual functional organization chart. Now and then, you can see the input-output type of diagrams for business functions. Some are presented in this book.

Another difficulty with the ordinary functional chart is the problem of showing two important concepts: (1) redundancy; and (2) time variations of the functional organization.

Examples have been given earlier to point out that many systems must have backup means to guard against unwanted malfunctions. Since the 1967 models, automobiles have had dual braking systems. If one goes out, the other still functions. This is an example of intentional duplication to insure reliable operation. In your home in the country, you keep a flashlight on hand, just in case. Sometimes the power fails. You may even have your own emergency generator. Such backups sometimes are not shown on a functional chart.

Moreover, time variations may be difficult to portray clearly. The functions of secretary and treasurer are quite different. Yet one person may act as the treasurer at one time and as the secretary at another. An accountant is essential. But he may not be needed all of the time. An outside C.P.A. may be called in to do the work.

To sum up, the idea of functional organization charts has a wide application. However, such a chart often does not give a complete picture. The incompleteness can be a serious drawback. Despite the drawbacks, a functional block diagram may be a good start for your system design thinking. You can check to see that everything shown is necessary—and that everything necessary is shown.

Functional Building Blocks

The concept of functional building blocks is probably the most important single idea in representing a system. Why? Because an existing or new system can be represented by a collection of functional building blocks.

There are only a few classes of such fundamental blocks. Any new system can be synthesized by choosing the proper members of these classes of blocks. A proper choice will perform all of the wanted functions. Once the blocks are chosen, all of the inputs, outputs, and interconnections can be worked out.

A system represented in terms of the basic blocks can be analyzed quite readily.

To build the actual system, each function can be performed by physical parts. The parts can then be assembled and interconnected to form the wanted system.

We have just said that any wanted system function can be performed using only a modest number of *kinds* of blocks. In fact, any possible block can be put into one of only five classes. Furthermore, of the five possible classes of blocks, you need remember only three.

Classes of Functional Blocks

What are the five classes of functional blocks? As shown in Fig. 2-8, they are as follows:

1. No input *or* output.
2. One input *or* one output.
3. One input *and* one output.
4. More than one input *and* one output *or vice versa*.
5. More than one input *and* more than one output.

No other classes are possible. The list is complete. Of the five classes, the first is the "no" case. It is only included to make the list complete. It has no apparent theoretical or practical importance in your system thinking.

Also, class 5 can be broken down into some combination of class 4 blocks.

Thus, in practical design, only classes 2, 3, and 4 are important.

A small number of additional concepts are now introduced. By their application, an extremely useful classification of functional building blocks is possible. These ideas are:

1. Inputs to or outputs from a block can be either energy or objects. As discussed in a later chapter, information may be carried by either energy, or objects, or both.
2. To change an object, energy must be expended. Work must be done on the object.
3. The blocks represent operations on the inputs to give wanted outputs.

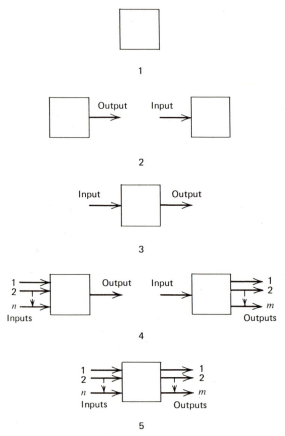

Figure 2-8.

4. *Inputs to a block* can be divided into *wanted* and *not wanted*. Similarly, *outputs from a block* can be divided into *wanted* and *not wanted*. Furthermore, any imperfect functioning of a block always results in an unwanted output—either unwanted energy, or object(s), or both.

These concepts allow us to draw Fig. 2-9.

The one input class has only one member: energy is applied to an object. Actually, this case is one often used as a short of shorthand. Of course, the energy applied to the object does not disappear (because of the law of conservation of energy). However, as a practical matter, the object may be thought of as an energy "sink." This idea is often useful in system design. For example, the atmosphere may be thought of as a heat sink for the hot gases from a boilerroom smokestack. You can think of many other examples.

The one output class has two members. The output may be either energy or an object, as the figure shows. Again, such representations are only a matter of convenience. You may want to show an energy source without worrying about where the energy actually comes from.

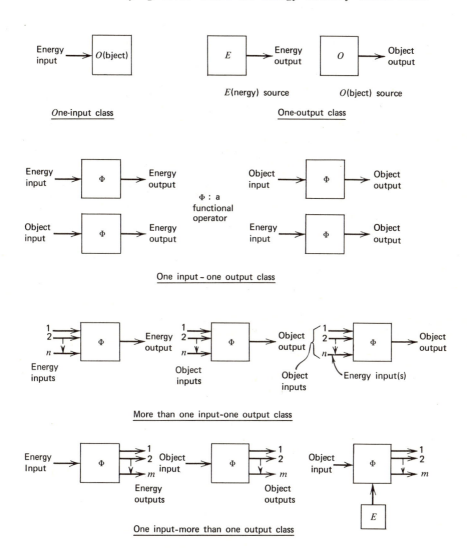

Figure 2-9.

For example, you don't show all of the details of the generation of the 60 Hz power that you need to make a television set work. You simply indicate that such a source is necessary by a power-source block. Similarly, if you have a machine into which you place an object for processing, you need not show all details of how the object actually got to the machine. For example, in designing a machine (system) to make machine screws, you need not worry about how the material (brass rod) was manufactured. A block can show the input objects to the system.

The one input and one output class has four members.

In the first case, energy goes into the system and energy comes out. A power transmission line is an example, if such effects as heat and corona losses are neglected. Energy is transmitted between two *locations in space*. The electric wave filters and equalizers widely used in broad-band carrier systems are also examples.

In the second case, an object goes in and a changed object comes out. Now everyone knows that ordinarily energy is required to change an object, and such a system requires at least two inputs (the object and the energy). There is one possible exception. Is energy required to change an object *in time?* Once you put a trunk down in the cellar, it just sits there. Yet it is continuously changing *location in time.*

In the third case, mass is changed directly into energy. This is possible and occurs in accordance with Einstein's equation $e = mc^2$. In such chemical reactions as the burning of coal or oil as a step in generating electricity, mass is conserved—not converted directly into energy.

The fourth case is also an example of the equivalence of energy and mass as expressed by Einstein's equation.

The more than one input and one output class is next. The energy-combination case is quite straightforward: the input energies are simply combined. At least in theory, many forms of energy can be combined. Dry cells can be connected together to give either higher voltage or more current. The cylinders in your car work together to produce more horsepower (which represents energy). Your high-fidelity stereophonic system may have two, four, six, even eight or ten loud speakers pouring acoustic energy into your living room.

The case of combining two or more objects to form a single object is a bit trickier. Why? Because to change an object in any way, you must expend energy on it. So, while this case is of theoretical importance, as a practical matter the situation represented by the third class is the correct one. Putting together objects is a common operation in manufacture. Therefore, the block representing such an operation is both useful and important in system design.

The one input and more than one output class also has three members.

The division of energy between two or more outputs is quite straight-forward. The transmission system and differential of your car divides the engine output between the two rear wheels. The 60 Hz power from a generating station may be divided among many thousands of sockets for TV's, toasters, washing machines, refrigerators, industrial motors, and the like. The division of an object into two or more objects always requires energy. The reason has been pointed out earlier. For this reason, this member of the class is of no importance from a practical standpoint. The practical situation is also shown. However, it has more than one input and also more than one output. It is shown to emphasize the point that an object cannot be changed without expending energy. Moreover, this block also represents the conversion of an object into energy such as occurs in fission and fusion reactions.

The blocks in Fig. 2-9 are idealized. In your design work, you will need them and you will use them, but always remember that they are idealized. You can use them for your preliminary studies and for setting up preliminary mathematical models.

One reason they are idealized is that nothing is ever perfect. There are always unwanted effects. Power sources are imperfect. Direct current generators have ripple; alternating current generators have harmonics. Input objects have imperfections of one sort or another: of size, shape, or composition. Imperfect inputs and imperfect system operation mean imperfect outputs. Thus, every system has some unwanted outputs.

What is the practical meaning of these statements?

Every system—no matter how simple—has more than one input; it always performs more than one function; and has more than one output. Hence, every system (even the simplest) must be represented by a diagram such as Fig. 2-10. There is always more than one input and more than one output.

In your design work you must check the wanted outputs. You must also be sure that the unwanted outputs are acceptable. Thus, design

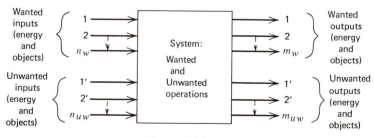

Figure 2-10.

work always has two fundamental aspects. You use the simplified
blocks for your first study. But be sure you do not neglect any unwanted input or any unwanted system operation. If you do, you may
be courting disaster.

The one output class can be further subdivided on the basis of the
function performed by the output. Thus, energy sources may merely
supply power; they may supply information carried by some form of
energy; or they merely contribute noise. Until we examine the question
of what noise is in more detail (in the next chapter), we shall merely
state that noise is unwanted energy or unwanted information. Thus,
the noise block is used to represent unwanted inputs. Similarly, objects
can be divided into three classes. Fig. 2-11 shows them for both
energy and object blocks.

Further subdivision is also possible for the one input and one output
class of Fig. 2-8. As shown on Fig. 2-12, this class can be divided
into those in which the block is *not* intended to produce a change in
the input and those in which it is. In the "no change" case, a medium
transports energy in space from one location to another. The energy
is not changed in any way. Similarly, a medium may transport an
object in space or time. The analogy with the energy case is not perfect,
however, because any change in an object requires the expenditure of
energy according to Newton's first law.

In the "change" case, the block alters some energy relations of the
input. In the literature, such a block is often called a "transducer." As
noted in the figure, there are a considerable number of different kinds
of relations that may be altered. Many of these have been extensively
studied for the case of electrical energy. In fact, many volumes have
been written about various possible changes represented by this func-

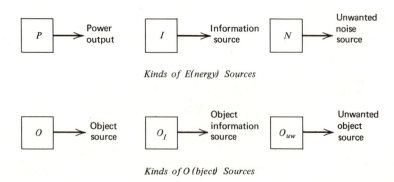

Kinds of E(nergy) Sources

Kinds of O (bject) Sources

Figure 2-11.

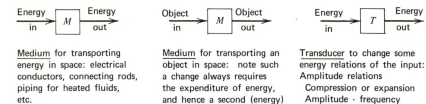

Medium for transporting energy in space: electrical conductors, connecting rods, piping for heated fluids, etc.

Medium for transporting an object in space: note such a change always requires a change always requires the expenditure of energy, and hence a second (energy) input.

Transducer to change some energy relations of the input:
Amplitude relations
  Compression or expansion
  Amplitude - frequency
  Amplitude - phase

Frequency relations
Phase relations
Space relations

Figure 2-12.

tion. A typical title is "Analysis and Synthesis of Electrical Networks." Some of the findings made by the students of electrical network theory have been carried over to mechanical networks. The telephone transmitter and receiver, phonograph pickups, and high quality loud speakers have all benefited. And these are only a few examples.

A special case of transducer function merits particular attention. It is represented in Fig. 2-13. The input energy is operated upon by the transducer (in this case called an "effector") so that it may be applied to change the object. Thus, this example includes the six simple machines: the lever, the block, the wheel and axle, the inclined plane, the screw, and the gear.

Fig. 2-14 shows some of the functions that may be performed by the more than one input and one output class. The block that does such operations may be called a "combiner." Input energies may be combined to form a single output. If they are later to be recovered and separated, the operation of combining is often called "multiplexing." Multiplexing is extensively used in the telephone plant. Many pairs of wires are combined in a single cable sheath. By using different frequency bands to carry different telephone conversations, a single channel

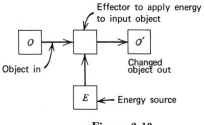

Figure 2-13.

may carry 1000 or more conversations simultaneously without interference. After traversing the channel, these separate conversations are recovered and sent on their way by a divider function shown in Fig. 2-15. The divider acts as a "demultiplexer."

The second kind of function performed by combining is computing: mathematical operations and logic operations. By arithmetic operations is meant the usual adding, subtracting, multiplying, and dividing. By logic operations is meant the performing of AND, OR, NOR, NAND, and similar operations. Modern computers using only these few different kinds of operations can solve extremely complex mathematical problems.

By a *comparison,* a combiner can determine whether the two inputs are identical or different. By a *discrimination,* it can tell whether two inputs are equal. If not, it can indicate which is the larger. One of the two inputs may be a reference value. Thus, measuring an input energy becomes a discrimination operation. The word "sensor" is often for blocks that perform the comparison operation. Sensing of an unknown quantity is an essential step in any measuring system.

Note carefully that a computation operation is not reversible. Take multiplication as an example. Suppose the product is 100. This might have resulted from inputs 1 and 100; or 2 and 50; or 4 and 25, etc. You can't tell what the input values were if you just know the product. You

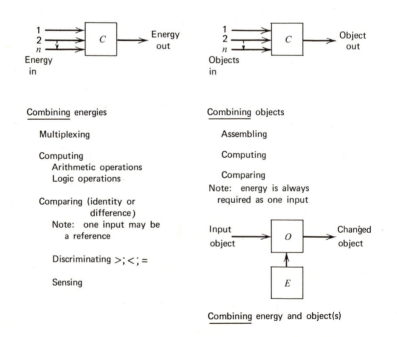

Figure 2-14.

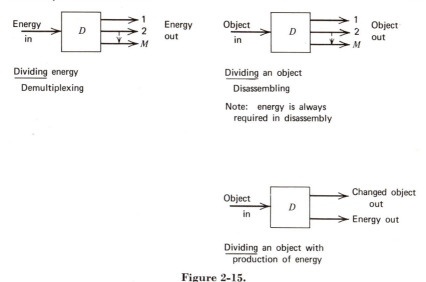

Dividing energy
Demultiplexing

Dividing an object
Disassembling

Note: energy is always
required in disassembly

Dividing an object with
production of energy

Figure 2-15.

can't even tell which input was the larger number. The situation is not exactly the same for comparison. But suppose you have an ordinary beam balance with two pans. You look at the indicator and see that the pointer indicates a balance. This is the output of the operation. But you have to know one more fact before you know the weight of the unknown object (one input). You have to know how many grams you put into the pan to obtain the balance (the other input). This means you have to know a reference value to measure and not merely compare.

If the inputs to the combiner consist of an object or objects and also of energy, then the operation of assembly can be performed. This is analogous to the multiplex using energies.

An object may be divided into two or more objects or disassembled by the application of energy. Finally, objects changed in some way by the application of energy may be used for computing or for comparison. The desk calculator is an example of a mechanical computing machine. It is either powered by pulling a lever or by an electric motor.

Blocks Considered as Operators

A functional block with one input and one output changes its input into an output. In other words, the output is the result of the operation of the block on the input.

A block with more than one input and one output combines the inputs according to some rule that determines the operation of the block. For an example, the block may simply add the inputs. This addition is the rule. Of course, such an addition may be described by mathematical formula. Also, a block with one input and more than one output shares its input among the outputs. The shares may be equal or follow some complex rule. But, in this case also, the block acts as an operator on the input.

This concept is sometimes called "modeling." The blocks are called "models" of the ideas or things that they represent. A more complete diagram made up of the blocks and their relations to each other is again a "model" of the subsystem or system that *it* represents.

Now, no such model is ever perfect. The representation is just that: a representation and not the real system. The word "rock" is not a "real" rock; a "block diagram" is not a "system."

In design work, you set up the simplest possible models that promise to yield the results you want. You neglect effects that you hope will not be important. For example, in designing a home radio, you may worry about the effects of heat on the operation but not about changes in barometric pressure.

Many models can be described mathematically. Where this is possible, the powers of mathematical thinking developed over the centuries can be used as a powerful tool in system design.

Despite the assistance from mathematics under the proper conditions, you must always beware of pitfalls and avoid them like the plague. Often, the actual operators can only be roughly approximated by your mathematical formulas. Too rough an approximation can lead you to wrong conclusions. You must always take precautions against such an unfortunate happening. Worse luck, you may find that evaluation of the possible error due to your approximation is quite an arduous task. But it must be done.

Connecting Blocks Together

Permissible arrangements

Functional building blocks may be connected together in only a few different ways.

One arrangement can be called "tandem" (Fig. 2-16). The amplifiers and cable pairs in a long telephone connection are in tandem. The cable loss further attenuates the already weak voice currents so they are restored by the amplifiers. In the coaxial cable systems used for long-haul telephony, amplifiers are necessary every mile or so. In a transcontinental connection, there are literally thousands.

A second basic way to arrange functional building blocks may be called "parallel." In some systems, the combiner and divider are required as shown dotted in the figure.

Fig. 2-16 also shows a *bridge* connection arrangement sometimes encountered in electrical networks. In this figure, block "A" is bridged

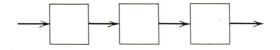

(a) Functional blocks in tandem.

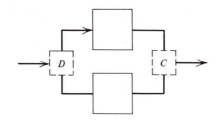

(b) Functional blocks in parallel.

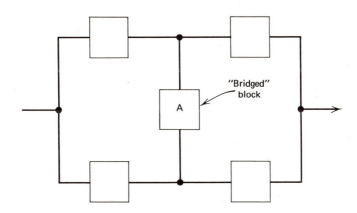

(c) "Bridge" arrangement of functional blocks.

Figure 2-16. (*a*) Functional blocks in tandem. (*b*) Functional blocks in parallel. (*c*) "Bridge" arrangement of functional blocks.

across the two parallel branches giving the arrangement its name. The Wheatstone bridge you used to measure electrical resistances in your physics laboratory is a familiar example.

Of course, a complex system may involve hundreds, thousands—even hundreds of thousands—of functional blocks arranged in many combinations of these few basic patterns.

At this point, notice particularly that we are discussing ways to represent the connections of functional building blocks. The blocks that we are diagramming only represent physical objects in the system when and if it is built. When the system is built, the physical objects must be arranged in space. The actual arrangement may bear little resemblance to that of the blocks on the functional block diagram. The functional block diagram is intended to clarify the operation of the system. The arrangement of the objects in the system may be dictated by such factors as size or appearance. These factors do not enter into the preparation of a block diagram.

Cautions to be observed in connecting blocks together

In your design work, the first step is to choose the blocks you need to do the operations you want to do. Then you arrange them so that the operations occur in the order you want them to.

So far, so good.

But you have to be careful.

For one thing, the output(s) of one block must match up with the input(s) of the next. For example, take an amplifier. A vacuum tube can give amplification. So can a transistor. But can you just replace a vacuum tube with a transistor? Almost certainly not. In your television set, the circuits are designed to use vacuum tubes—not transistors. Either device can act as an amplifier, but to change over from one to the other means a revision in the circuit design.

As a more complicated case, suppose you are working on a rocket, using electrical and hydraulic components. At the interface where the two kinds of components meet, they must be made compatible. Furthermore, an energy conversion is almost certainly necessary. At such interfaces, great care must be taken to avoid an unworkable condition. A farfetched example? Not at all. In complex systems made by different companies in different factories thousands of miles apart, the chances of such errors are not negligible. Constant vigil is necessary. To work together properly, parts must fit together. Plugs and jacks must mate.

Don't forget the interconnections between blocks: the blocks that transport energy or objects from one place to another. In the actual system, they must be carefully tailored to perform their tasks.

To sum up, a good functional block diagram is just a beginning. You must find out how you can do each required function—and how to make sure everything will work together.

Summary

Without a well-thought-out functional organization, the design of complex systems would be impossible.

Two very useful and widely used representations of systems organizations are:

1. Diagrams showing the functions to be performed.
2. Diagrams showing the blocks necessary to perform each function and the interconnections between the blocks.

The first representation gives an overall picture of what is to be done. The second gives much more information about how the system works. Both have a place in design.

The possible *kinds* of functional blocks can be and have been tabulated. Furthermore, the list is complete. Any system must be made up of some combination of these few kinds of blocks. Of course, in real life, you may be able to buy collections of devices that perform each function.

At least in theory, the operation of all the kinds of functional blocks can be represented mathematically. Such a representation can be extremely useful. But the limitations of mathematical knowledge today make experimental checks of system performance essential. Often, such a check turns up surprises.

The possible *kinds* of interconnections of functional blocks are few in number. Also, this list is complete.

3

Some Information Theory
Concepts in Design

Design Involves Information Processing

Design is an information-processing activity. It involves manipulating signs and symbols either in fact or in the imagination. For example, you may write an equation and manipulate the symbols in accordance with the rules of algebra or symbolic logic. You may draw a geometrical figure as a help in proving a theorem. You may draw a block diagram of a new organization, or of a new system. Or, you may lean back in your chair and think about such concepts, and not bother to put them on paper until later—if at all.

In these examples, and in many more that you can think of, you take some input information, process it, and come up with output information. What you actually do is to transform input information into output information.

Thus there are three questions to be answered:

1. What is information?
2. What manipulations are possible on input information?
3. What are the limitations on information processing?

When you have this necessary background, then we can draw the kinds of information inputs, outputs, and processes that are used in each step in a design procedure.

Definitions

Modern "information theory" has made many contributions to understanding important aspects pertinent to our present discussion. Much of it is not directly applicable but can throw much light on the nature of the problems raised by the three questions.

One value of modern information theory is the fact that it carefully defines the words it uses. When precise definitions are used, it is possible to put some basic concepts in mathematical form. This is an enormous advantage in making the concepts meaningful.

Information and communication*

Information-in-general can add to a representation of what is known, believed, or alleged to be true. Thus information can increase your knowledge. Actually, it may or may not do so. You may not need the information, or you may not know it exists, or you may just not pay attention to it. After all, some students do nod in the classroom.

An *information source* generates information. Obviously, a person speaking can be a source. He might be giving a lecture or a talk; he may be the next-door neighbor with a bit of gossip you did not know. Or, an information source may be a book, a photograph, a newspaper, or any one of a myriad of other things. A piece of rock you pick up may tell you a story of the geology of the mountain in front of you.

Communication of information is the reproduction of information from a source at some other location or locations. The reproduction may be either exact or approximate. Radio, telephone, and telegraph are examples of methods used to communicate information. But you can also communicate information by sending a letter. A particular swatch of material may tell someone else exactly what color you wish to get in painting a room.

Signs and symbols

Communication cannot take place without a *system of signs*. A *sign* is defined as follows:

1. A written mark or marks conventionally used for a word or a phrase. For example, combinations of the letters of the word "sign" itself.

2. A natural or conventional motion or gesture used instead of words to convey information: a wave of "goodbye."

3. A thing regarded by general consent as naturally typifying, representing, or recalling something by possession of analogous qualities or by association in fact or thought. Such a thing may also be called

---

* *Information, Computers, and System Design,* by Ira G. Wilson and Marthann E. Wilson, Wiley, New York, 1965, is a book in Wiley's System Engineering and Analysis Series. The definitions and discussion immediately following are derived from material in Chapter 3 of the book.

a *symbol*. The audible dots and dashes of a telegrapher are one example. Uncle Sam, Santa Claus, a Red Cross, and a white flag may also be regarded as symbols. Our modern culture is full of symbols. In fact, by strict interpretation of the definition, *any word is a symbol*. Thus, in the English language, the word "earth" is intended to convey a particular idea. Obviously, the word is *not* the earth. It merely represents or recalls it to your mind. Note that the Germans use the word *Erde* to convey the concept of "earth."

Signals

A *signal* is the manifestation of some physical phenomena by which information is transmitted. Thus, by definition, signals have an information content simply because they are conveying information. Electrical, light, and sound waves are frequently used for the purpose.

Signals are the only physical manifestation of any information. Any operations on information can only involve operations on the physical phenomena carrying the message. As a human being, you can receive information through your senses. And by far the most important ways in which you receive information are via your eyes and ears. In other words, you get almost all of your information from the light waves entering your eyes and the sound waves impinging upon your ears. Also, you convey information by the words that you speak and by the movements of the muscles of your body (particularly those muscles that control your hands).

Messages

For the present purpose, a *message* is an ordered selection from an agreed set of signs intended to convey information. Since a message is a selected amount of information, a single word can be an entire message. In the average household, a frequent message is "no" or "don't." A telegraph message is often only ten words. But according to the definition, there is no limit on the number of words. A paragraph, a chapter, even a book or a whole encyclopedia may, on occasion, be considered to be a message. Thus, the number of messages does not necessarily equal the number of signs or words in the message. To repeat: by definition, a message is simply a selected amount of information.

Codes and coding

A *code* is defined as a set of rules on any subject; a system of military or navigational signals. A *coding* is a transformation or mapping. Coding or encoding is the application of transformation rules to a signal or message conveyed by symbols or signs.

Decoding is the inverse operation.

Writing and reading are examples of the coding and decoding of information. You write down the signs (for example, in the English language) corresponding to the symbols intended to convey particular information. You decode information in a book when you try to recover what the author intended. As pointed out in the Wilson's book, the users of a code must both understand the intended correspondence between the elements of the code and the things denoted by them. In telegraphy, your words are broken down into letters and each letter is coded so it can be transmitted by a set of dots and dashes. Semanticists called this understanding the "meaning" of the symbols or signs. Thus:

1. All oral and written languages are codes.
2. Telegraph codes and secret codes and ciphers are codes of codes.

Ordering in Space and Time

Ordering is a very important factor in the structure of signals. Letters are ordered to spell a word; words are ordered in sentences; and sentences are ordered in paragraphs.

Rearrangement of the letters of a word or words is called anagramming. Even a minor rearrangement may destroy the meaning completely. As a simple example, by anagramming you can turn "dear" to "read". The letters are the same; only the order is changed; and the meaning is completely changed.

Ordering is also important in design. For example, you can't process information that you do not have. And you can't select the best alternative proposals until after the proposals are formulated. These are only two examples. Remember that the importance of ordering in design effort can't be overemphasized.

In practical thinking about a problem, you seldom are able to take the steps in your development (say 1, 2, 3, 4, etc.), one right after another. Instead, you may work through 2 and 3; there you suddenly gain a new insight into the problem (more information). So you go back to step 2 and repeat it, taking into account what you learned: you "recycle." But after repeating step 2, you go on to step 3, then to 4, and so on *in order*. You can't start with 4, and go backward 3, 2, 1.

Furthermore, the steps are *ordered* no matter how often you recycle, or where you enter the ordered sequence when you begin a recycle operation.

In one sense, design work is not a formal, rigid process. You recycle, perhaps again and again. You revise, revise, revise! Yet there is an underlying order that you must understand in planning your thinking.

Measures of Amount of Information: The Unexpectedness Concept

An *amount of information* may be measured in terms of at least three different concepts. However, one of these—the *unexpectedness concept*—will suffice for our present discussion.

To measure an amount of information by using the *unexpectedness concept*, you put yourself in the position of the receiver of the information (or possible information). The message or signals containing the information may be either discreet or continuous functions of either space or time or both. The same unit may be used for any of these cases. Then a particular amount of information from the source is considered as *selected* from a set of possible alternative amounts of information. The important word is "selected." The amount of information is the *unexpectedness* or *improbability* of that particular information being selected, without paying any attention to the actual structure. According to the unexpectedness concept, if you know the information before you receive the signal, then you gain no information. The information content is zero. To illustrate this important point: suppose someone tells you $2 + 2 = 4$. If you are over 6 years old, you already *know* $2 + 2 = 4$. You expected the answer; you got no information. Again, suppose your telephone rings at 3 A.M.; you answer and hear, "The house next door to you is on fire." You certainly didn't expect these improbable words in the middle of the night: so you got information you did not have before the call woke you up.

Hartley showed that when a particular amount of information was selected from a set of possible alternative amounts, the logical choice for a measure of information is logarithmic.

If the base of the logarithms is 2, the unit for expressing a quantity of information is called a "binary digit" or simply "bit." If the base is 10, the unit is called a decimal digit (or sometimes a "decit").

The information content of a message is the number of bits conveyed by the message.

The number of bits per unit time (frequently per second) is called the information rate. Similarly the number of bits per unit area of space (such as per square centimeter) is called the information density.

With more than one source of information, the amounts from the sources are added.

What Is Noise?

Suppose you are at a large cocktail party. At a particular time, ten (or more) people may be talking at once. But you want to pay attention only to what your partner of the moment is saying. Under these circumstances, the other nine (or more) people who are talking are a nuisance; to you, they are contributing nothing but "noise." You are being bombarded by multiple signals. You want only one of these— your partner's. In information theory terminology, your situation is described as "a signal corrupted by noise." And *noise always causes a loss in the amount of received information.*

In the cocktail situation, the signal from your partner is an acoustic wave. So are the signals from all of the other people speaking. The distinction between the wanted signal from your partner and the noise is simply their relative importance to you. At the cocktail party, by carefully turning your head and concentrating your attention, you may be able to filter out much of the interference. Exactly why this particular filtering is possible is not completely understood. A radio set may not be able to distinguish between incoming signals. For example, "static" may interfere with your listening; and you cannot always tune out static interference.

The concept of noise is useful in many areas. For example, a fingerprint that obliterates an important dimension on a drawing or mars a photograph may also be called "noise." By analogy, the interjection of unwanted ideas in a conference is also "noise." You can think of many other examples.

An Important Theorem about Information Processing

Imagine any kind of information-processing device. The device may have one or more inputs and one or more outputs.

Claude E. Shannon, then of the Bell Telephone Laboratories, proved: *over a long time interval, the amount of information from the output of any device is always equal to or less than the amount of input information. It is never more.* This is an extremely important concept. Stated differently, Shannon's theorem says that over a long time *no device can increase the quantity of information put into it.* Perfect reproduction and complete freedom from noise is impossible.

All information processing by machines is subject to the basic law. In simple words, these mathematical developments show what can and cannot be done in any possible information processing operation.

By his basic contribution, Shannon showed that the total information output of any machine cannot exceed the total information input. In practical machines, the amount of output information must always be less than the amount of input information. It may be much less; in fact, it may be zero. Paper-shredding machines are made to chop up secret documents so that their information content is destroyed.

Shannon's remarkable theorem shows that there is no "law of conservation of information." In physics, you are taught that energy is conserved. So is momentum. But at least in any sort of machine, information is *not* conserved.

Scientists familiar with the theorem accept it for any mechanical, electrical, or other type of device. However, when it comes to the human brain, some scientists question whether it holds true. The question is still unsettled. But then information theory itself is only about 30 years old. The flights of fancy of Shelley and Keats have been with us much longer—not to mention the Iliad and Odyssey. Quite possibly, human beings may have powers beyond those of any mere machine. And then again, we may not. In other words, can you know or imagine something in the absence of any prior knowledge? Many people have imagined what Martians look like, although they never have seen one and may never see one. From what sort of information do they conjure up their picture of the Martians? A difficult question. And one even more difficult to answer.

For a long time, a great deal of argument took place about the question of conservation of energy in the human body. Today, it is generally agreed that the human body does conserve energy. Possibly Shannon's theorem holds true for our brains.

Information Theory and Design

What has information theory to do with design? A great deal.

Design is a step-by-step procedure (with recycling). For a particular step, certain input information is necessary. Furthermore, a processing program must be available. The input is processed to give an output. In turn, the output becomes part of the input for the next step.

As a consequence of these facts, an ordering of the steps in the design procedure is unavoidable. This is an important finding. They also explain the reason for and the importance of recycling and revision as new information becomes available as the design work progresses.

As a specific example, suppose you are engaged in a design and you find you are lacking some information. To be specific, take the cost of some part that you wish to incorporate. To get a meaningful total cost, you must either obtain good information about the missing item or just guess at it. If you guess, and your guess is wrong, the outcome may be disastrous. The results of such poor estimates can be seen in the columns of our daily newspapers. We find that almost every item of military and space hardware costs half again as much, twice as much, even ten times as much as was estimated in the beginning. In our private lives, if we allotted $3000 of the family budget for a new automobile and the cost turned out to be $30,000, the results could be disastrous. Perhaps we are seeing some of this in our national affairs.

In any case, the lack of necessary information can be disastrous. And information theory states quite clearly that the total amount of output information from any machine can never exceed the total amount of input information.

4

Information Handling

The Basic Sequence of Information Handling

Several concepts presented in earlier chapters are used here to study a basic procedure in system design: information processing. The functional building blocks described in Chapter 2 will be applied to the problem. In doing this, some ideas of information theory discussed in Chapter 3 give invaluable clues about necessary inputs to the functional blocks. And the objectives and goals (the MUSTS and MUST NOTS) of Chapter 1 also appear as required inputs.

Fig. 4-1 shows the basic sequence. First, you try to acquire or retrieve information from an information source or set of sources. Hopefully, as a result of your efforts, some pertinent items are found. As pointed out earlier, you can't process information that you don't have. You don't act rationally before "getting the facts."

Next, the acquired or retrieved information is "processed." In the human case, the processing may take the form of "thinking things over." But "thinking" is a rather vague term, and later we shall give it more definite meaning.

Again, hopefully, results ensue from the processing. These results form the basis of some human action which is the output of the chain of steps.

Notice that in Fig. 4-1 the information actually used is only a small portion of that available from the sources. The available information may be thought of as flowing down the branches of a tree to the trunk located at the input to the processing. Similarly, the output of the processing diverges to a number of possible actions. Usually, only one is chosen and put into effect.

A number of important remarks about the steps in information processing can be made by studying Fig. 4-1.

First, the steps are well-ordered in time. By "well-ordered" we mean that there is a first step that precedes all others. You must acquire

60

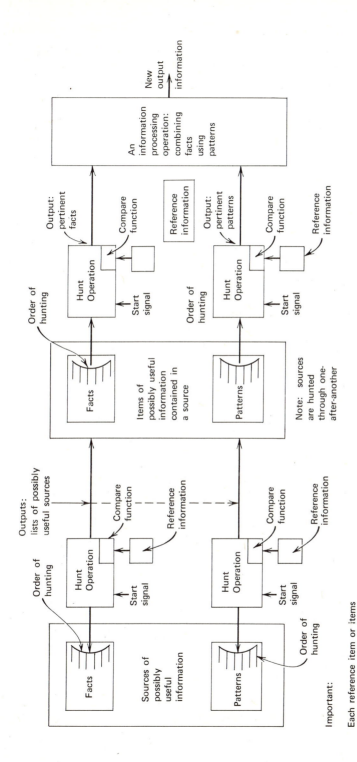

New output information

An information processing operation: combining facts using patterns

Output: pertinent facts

Compare function

Reference information

Output: pertinent patterns

Compare function

Reference information

Hunt Operation

Start signal

Hunt Operation

Start signal

Order of hunting

Order of hunting

Facts

Patterns

Items of possibly useful information contained in a source

Note: sources are hunted through one-after-another

Outputs: lists of possibly useful sources

Compare function

Reference information

Compare function

Reference information

Hunt Operation

Start signal

Hunt Operation

Start signal

Order of hunting

Order of hunting

Facts

Patterns

Sources of possibly useful information

Important:

Each reference item or items used in the hunting operations is the result of an earlier information acquisition and processing procedure.

Figure 4-1.

information before you can process it. You must process it before you act on it.

The order is not reversible. You cannot act before you process; you cannot process information before you have it.

Only a small number of kinds of activities are shown: acquisition or retrieval of information; processing; and taking action.

The input tree and the output tree are finite. The number of possible sources of information may be very large. Yet the number is still finite. Even the largest library has only a few million books. The number of possible actions may be large; yet it is finite. Look at all the books? Or certain titles or authors? Or just one book? The fact that the trees are finite may be of small comfort. For example, take a game of chess. The number of kinds of first move (actions) for the first player is very small; likewise, for the second player. Similarly, for the second move, third move, and so on. Yet the tree has so many possible branches that the number is almost unimaginably great—yet, it is not infinite.

The three kinds of blocks in Fig. 4-1 can be broken down into smaller functional units. This further breakdown will be carried out presently.

Earlier, the fundamental building blocks of systems have been given. These blocks will now be applied to the functions performed in the information handling sequence.

Storing and Retrieving (or Acquiring) Information

As a designer, you will often be required to retrieve or acquire needed information. To get it, you must search some kind of information storage: a human brain, a book, or one of many other kinds of storage media.

For you to be able to get the information you want, it must have been stored previously. Information that is not stored cannot be retrieved or acquired. If a book that you want is not in the library, you cannot look at it or draw it out. Furthermore, if the book you need has not yet been written, no search will ever find it. These are blunt statements. But people *do* go looking hopefully for information in places where it cannot possibly be stored. They ask advice of people who have no pertinent experience. Our newspapers are full of stories about committees of nonexperts that are set up to find answers to problems about which they know little or nothing. It is obvious that a book must be put on the library shelves before it can be drawn out for consultation. Similarly, information must be stored in a human brain before it can be retrieved and used.

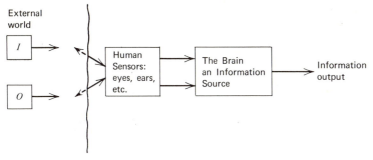

Figure 4-2.

The possible sources of information you may consult includes people, graphic material such as books, and objects such as machines and computers.

In Fig. 4-2, the brain is represented schematically as the source of information. In turn, it acquires information through the sense organs of the body from the external world. From the observation of the external world and by using imagination or inspiration, you sometimes can get a thought (idea) of a new want, a new need, or a problem of your own or someone else's. All of these kinds of ideas result from thinking—perhaps wishful thinking.

Insofar as we know today, the human brain conceives all of our wants, needs, and problems. In this area, a computer is no substitute for the brain. The human brain both sets the objectives and the requirements to be satisfied. It decides the functions to be performed. It chooses the MUSTS and the WANTS that the new system must meet. Also, it sets up the MUST NOTS and the DON'T WANTS.

In other words, it sets up the goals and the rules of the game. Of course, human desires cannot circumvent natural laws (such as the law of conservation of energy). Perpetual motion is still impossible. But when you decide you want a new automobile, you can put your special requirements upon the car that you want.

The importance of patterns in processing facts and other information has been pointed out earlier. Many of these patterns are taught us in elementary school. You learn how to add and subtract. You learn the multiplication table and how to divide. Later on, you study more advanced mathematics. You learn the rules of logic and their application to clear thinking. In fact, much of our "thinking" is the application of the patterns of thought taught us in school to the problems of everyday life.

Despite its enormous capacity, the human brain does have limitations.

For this reason, we store much useful information in books, in tables, in photographs, and other written and printed forms. The telephone book is only one example. The Great Books of the Western World contain excerpts from the wisdom of the philosophers and thinkers who have preceded us. The volume of available material is so enormous that it would be useless to try to retain it in the brain. Instead, you need the ability to locate the right source and find the exact item of information when you want it.

The Information Storage Process

Information can only be stored in some object in space. The object may be your brain, a book, a piece of paper, or one of many other things. As pointed out earlier, a piece of rock picked up by the roadside may tell a geologist much about its origin and the changes that it has undergone. Information may be stored on blank paper by writing, typing, or printing. But always, information is stored in or on an object in space.

After information is once put in storage—no matter what the medium or manner of storage—it always decreases with time. For example, your memory fails; pages in a book become yellow and torn; and the rock is worn down as it tumbles along a creek bed. Deterioration may be fast or it may be slow. But it always occurs.

We should not be surprised at the deterioration of stored information with time. It is an example of Shannon's theorem that any device or machine can only deliver the same amount of or less output information than is put into it. It can never deliver more. We humans are particularly fallible because many things that we see and hear are not remembered at all. Only the unusual or the repeated things seem to force their way into our memory so that we can recall them later.

The development of the computer and the "information explosion" that we read so much about has made us conscious of the large amounts of information available for storage. Theoretically, all of the possible moves and countermoves in a game of chess could be stored in a computer. Then after the first move was made, the outcome would follow as a matter of course. But the amount of storage space required for all possible moves is so enormous that it is beyond any imaginable computer capacity. Simply because of its limited memory capacity, a computer cannot play as good a game of chess as a skillful human opponent.

There is yet another side to the information storage problem. To store a large volume of information may require a large volume in space. Books take up shelf space in libraries; letters and design drawings take up file space in offices. The volume of stored information is becoming so large that methods of reducing the required space are used in many places. Microfilming of books and documents is only one example.

The function of storage of information can be put into system format. In some mysterious way, we "memorize" facts almost unconsciously. Also, we give little thought to filing a letter. But if we break down these operations into their fundamental functions, then it becomes clear how complex they really are.

To begin with, we must have:

1. A *set of storage locations* suitable for the items to be stored. These locations must exist in space (or possibly in space-time). For books, they may be the "stacks." It is impossible to store information in time without the existence of some object in space.

2. A *set of addresses* for the locations in the store. These addresses may be thought of as tags for the possible locations: the number for each shelf in the library. Since the locations are in space, the tags must also exist in space.

The set of storage locations might be thought of being like a file drawer in your office. The set of addresses are the alphabetic dividers used so letters can be filed and more easily retrieved.

3. A *program* to reach a preselected empty store; or an order of examination to select an empty store and method to make sure that it is empty. You must know how to get to a particular shelf where you want to put a book.

Unless you want to destroy an item already in a particular storage location, you must make sure that it is empty before you put in your new information. Furthermore, there must be some method of reaching any storage spot that you want. This is the program.

4. A *change* must be made in the state of the selected location. You put the book you want to store in its place on the selected shelf. Note that any change in the state requires an expenditure of energy.

When you put a letter in a file you change the state at that location. To insert the letter, your muscles must be energized. This simple principle that expenditure of energy is always required applies to magnetic cores in a computer; to putting a book on a shelf in a library; and to the human brain.

5. The exact *address* of the selected location in which the information was stored. A particular book has its own place on the shelf in a well-run library.

6. A *catalog, record,* or *memory of the address* so that the stored

information can be retrieved. In a library, a book has one or more cards describing it and its location in a "catalog."

Note that two entries are necessary here: the locations of storage places *and* the items of information stored in them. When you want a particular item you consult this list. Then you go to the location and retrieve it. In a file drawer, all letters from company A are filed in the A location; those from company B in the B location; and so on. You or your secretary remember the locations in this simple way. The situation is much more complex in a computer. Items are not filed alphabetically—perhaps not even systematically. So a separate list of locations and items in them must be kept within the computer memory so that wanted information can be retrieved when it is needed.

7. To perform these six functions, *signals* of some kind are necessary. First of all, a signal is necessary to start the operation. A signal is necessary to begin the selection function to find a suitable storage address. This signal may have to contain information about the amount of storage space required. Usually, there is enough space in the file drawer to add one more letter. But a library shelf may not have enough space to add one more thick volume. If the space is not available, then some further action is necessary before the book can be put in its final place. Also, signals must be generated to store the address information telling where the item was placed. In a library, a catalog contains such location information—as well as other information about what has been stored on the shelves. Signals are necessary to direct you to records telling the location of where the wanted item is stored and also to read the recorded information. Also, control signals are necessary for the energy supply to store the item. These signals must start and stop the energy supply as well as direct it in space.

We human beings store information in our brains. Presumably, to do this we perform all of the functions enumerated above. And to perform these functions, the signals just described appear to be necessary. More discussion on this point occurs in the final chapter.

The Information Retrieval Process

The words "information retrieval" mean exactly what they say: the recovery of information previously stored. *Retrieval is not discovery.* Discovery requires the appreciation of some unusual or unexpected condition or fact. For example, you say that Dr. Fleming discovered penicillin. A long-lost copy of Leonardo da Vinci's works was "discovered" a few years ago in a Spanish library. There was no reference to the manuscript in the library catalog and the discoverer came across it quite unexpectedly when he was looking for some other information. In information retrieval, you are trying to get pertinent information known to be (or believed to be) possessed by a certain person or contained in a particular file.

The important point for your to remember is that information retrieval or acquisition always requires two steps:

1. Selection of a possible set of information sources believed likely to contain pertinent information: perhaps many books, a few books, or one book.

2. Selection of the particular item or items of information you want.

Selection of only pertinent items of information is important. You do not—or should not—want a large amount of useless information. For this reason, careful selection among possible sources, and rapid search of these sources should be your aim.

In information retrieval, the list of requirements that must be met in acquiring an item is even longer than the list given above for storage. You met some of them in the list of requirements for storage. This is quite natural since we are talking about storage *and* retrieval of information. An item not stored, stored incorrectly, or stored without a record either cannot be retrieved or only retrieved by relatively inefficient search. A missing or misfiled letter that you need is a good example. But it is not a simple matter conceptually to find even a correctly-filed book. The reasons are as follows:

1. You need a *set of locations* of information. *Each location must have a corresponding address.* These are the library "stacks."

2. You need a *selection process* to determine which locations contain pertinent items of information. You consult the catalog for likely books.

3. To make selection possible, *access* must be gained *to wanted store locations.* You or someone else must be able to go to any shelf in the stacks.

4. A *signal* (such as a query) must be given to the location to indicate that information is wanted.

5. You need a *strategy or program* for examining the locations (for example, of wanted books). You may set up a particular order of examination as a strategy, or you may query them at random. These are only two possible kinds of programs.

6. You must *signal to a chosen location* that you are querying it. In some simple cases of information retrieval, all that is necessary is to look at or ask whether any information is in a particular location. But in more complicated cases, where you want a particular item, you must

put your query in very specific form. Thus to get the telephone number of a particular person, you must know exactly how his name is spelled and possibly even his address.

7. In many information retrieval systems, you must have available the *address of a wanted item* and tell a person or machine you are querying where and, in some cases, when to look for it. In a library, you get the address from the catalog card for the book.

8. You must have *access to the stored item* through some form of connection to it. As examples, the connection may be electrical, mechanical, or by light beam. You must be able to reach the book you want.

9. You must have a *sensing arrangement* to tell whether a selected location is empty, or occupied and, if occupied, whether it is suitable. In other words, the selected location must be evaluated. The evaluation may be done in more than one step. For example, when you go to the library, you get from the catalog the location of books that may contain pertinent information. You glance through the table of contents of each book and put aside those which may justify a further look and return the rest as not pertinent. The evaluation function always requires a rule or rules which determine whether to accept or reject a particular source. When you go to the stack, you look and see whether the book you want is there.

10. Quite a few *signals* are required to perform the function of acquiring, selecting and evaluating possible sources of information. Only two will be mentioned here: the start signal, which often takes the form of a query; and the stop signal, which ends the search.

11. An *energy source or sources:* power supplies. Each functional operation and necessary signal to control the operation requires the expenditure of energy. These energy sources may be of different forms—and often are. As just one simple example, when you reach for a book on a library shelf you are using muscular energy. When you look at the table of contents, the information is carried by light energy.

After one or more possible sources of pertinent information are found, the sequence of steps must be repeated to find out whether each source really contains any useful information. In some searches, you may look at each source as it is located, search it for the particular item you want, and then retain or discard it, depending on the results. In other searches, you may locate a number of sources, screen them quickly, and then put some of them aside for a more careful examination later. The particular procedure that is used may depend on the circumstances or your personal preferences.

At this point, it is easy to see how you find information in a library by going through the sequence of operations step by step. Libraries file their books on their shelves by categories. You can find the address of any book and also an idea of its contents by consulting the card catalog. Things are organized so that searches are relatively easy. Of course, you may miss a possible source because a title is misleading, or the information on the catalog card is incomplete.

Block Diagrams for Storage and Retrieval

The processes for information storage and retrieval can be represented in block diagram form. Retrieval is the more complex of the two, and will be examined in detail. Perhaps, as an exercise, you may wish to put together a block diagram for information storage using the functional blocks presented in Chapter 2.

The retrieval process always starts with a query for wanted information. Furthermore, the query must be phrased in a language understood by the person or machine being questioned. An often-quoted example of the difficulties in this simple statement is the phrase "airplane production." To a human being, a query phrased "aircraft manufacture" very likely would cause no difficulty. But computer-based systems of information retrieval might well have some problems with even such a simple change of wording—although the meaning of the two phrases is almost identical.

As is common in making block diagrams, we can start with Fig. 4-1. You first search for likely sources of useful information. Based on the results you obtain, you search the individual sources and obtain more results.

Remember that when you search for information, you almost always have to search for, process, select, and evaluate possible references that will enable you to separate the wheat from the chaff. And this is no easy task. If your references are not well selected, you may receive an avalanche of information—or none at all. Worse yet, you may receive a mass of irrelevant matter. None of these outcomes is desirable. Therefore, set up your references with care.

Fig. 4-3 represents a more detailed block diagram of the search and evaluation processes. As shown in the figure, the inputs are searched one at a time and each output from the search is then evaluated. Also more than one reference may be used.

The possible outcomes of any search are given below:

1. One complete (successful) match with the references.
2. More than one complete match.

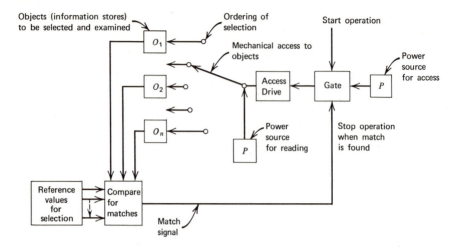

Figure 4-3.

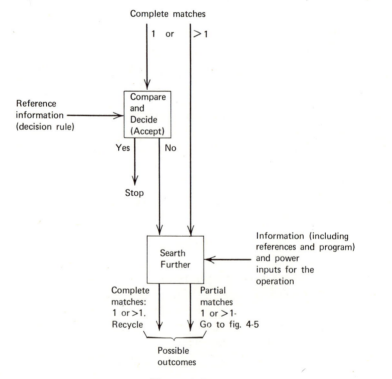

Figure 4-4.

3. One or more than one partial (not competely adequate) match.
4. No partial matches.

If there is only one complete match, then the situation can be represented as in Fig. 4-4. Here the reference is the set of decision rules that determine the adequacy of the match. Possible actions are to stop the search or, alternatively, to continue to look further. You either seek more sources or you examine those that you have retrieved more carefully.

With more than one complete match, the objective is to choose the best of the alternatives. First, the several possibilities are put together in an ordered list. Then, each item in the list is selected and reevaluated more carefully. Finally, the best alternative is chosen.

If only partial matches result from the search, then the situation may be represented by Fig. 4-5. Each partial match (if there is more than one) is selected and compared with an appropriate decision rule. The possible outcomes are shown together with some suggested actions.

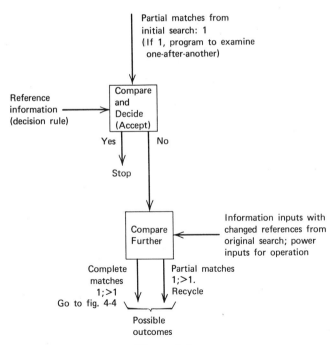

Figure 4-5.

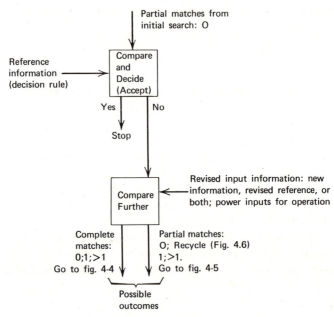

Figure 4-6.

Finally, Fig. 4-6 shows the possibilities if the initial search turns up no partial matches—an unhappy situation.

Notice that the list of outcomes is logically complete: there are no other possibilities. In a search for sources or items of information, the possible outcomes of an evaluation of a particular item may also be tabulated as follows:

1. *Intolerable shortcomings:* does not meet all the MUST requirements. This is fatal, and the source or item is immediately rejected.

2. *Quite undesirable shortcomings.* In this case, you may either reject the item or take some positive action to modify or eliminate the cause of trouble.

3. A *minor shortcoming* or an important one with a low probability of occurrence, or both. If the shortcoming is minor, perhaps you can eliminate it, provide against it, or simply minimize it. Or if the shortcoming is more important but not very likely, you may choose to ignore it and take a calculated risk.

4. *All of your* MUSTS *and* WANTS *are met.* This is a happy ending indeed to that particular search. You have found the exact item of information you want!

Information Processing

Now that you have acquired or retrieved some items of information, what next? You process them in an appropriate way.

Only four different *kinds* of processing are available to you.

From the discussion of possible functional building blocks in Chapter 2, you have:

1. A one-input, one-output block.
2. A one-input and more-than-one output block.
3. A more-than-one input and one-output block.

    (a) An object input and an energy input to change the object.

    (b) More than one information input.

Of these four, *1* and *3a* are of much less importance in the present context than *2* and *3b*. The two less important possibilities will be disposed of first.

Fig. 4-7 represents information going into a process and emerging changed. Such a processing block is sometimes called a transducer. One such process conveys the information from one point in space to another. Actually, in some telephone systems, transducers are used to compress or expand the electric waves carrying the information. Also, other tranducers such as wave filters selectively distort the input signals. But since we are not considering the design of telephone systems here, these possibilities can be put aside. Thus the transducer process is of only secondary importance in the present context. You talk to your colleagues without ever thinking about the air path that conveys your sound waves. You use the telephone as a tool; you pay little attention to how it works. Yet these are examples of transducers with which you are familiar and which you use every day.

Fig. 4-8 shows the other example that is of only secondary importance here: the processing of objects so that they may convey wanted information. A common example is given by a piece of blank paper as the "object in." The "information in" is what you write on the paper. The "object out" is the paper with the writing on it. You can

Figure 4-7.

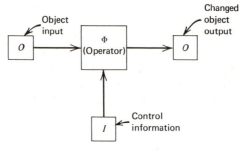

Figure 4-8.

think of many other examples: typed manuscripts; mechanical drawings; punched cards; specifications; and instructions. In all these instances, some object is changed to convey information. The information
may come from a human being or from a machine such as a computer.

Fig. 4-9 represents the process of *division*. Sorting is an important
example. In communication systems, sorting sometimes is called demultiplexing. For example, in your television set, the picture information
is separated from the sound information. The video information is presented on the face of the picture tube; the sound comes out of the
loud speaker.

In system design, information-bearing documents may be sorted into
appropriate piles or pigeon holes. In accounting, expenses are charged
to appropriate accounts.

Erasing to decrease the amount of information present in the input
is a special case of sorting. You recall that information does not follow
a conservation law. You can simply erase it or throw it away.

Notice that in Fig. 4-9 the division of information requires some sort
of control information to tell where each input item is to go. This
control information may be wired in, as in your television set. In a

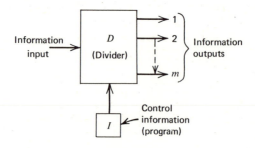

Figure 4-9.

card sorter, it may depend on a special program. Or it may depend on a human brain, as in editing a manuscript.

Proper sorting of information into categories is often used in system design. It can be a very useful tool. It can bring out patterns in raw data that would otherwise go unnoticed. And the right kind of sort is often essential to make heads or tails of a great mass of data. But the right kind of sort means the right control information. And to get the right control information, you must go back and acquire possible kinds of control, and select the most appropriate one. Thus we are back again to the information acquisition or retrieval problem.

Allotting is another form of division. In starting a new design, management must allot resources to the project. For example, people must be made available, either from those already on the payroll or by hiring. Designers also perform allotting; for instance, you are told that a new system must not weigh more than 100 pounds. So, based on your judgment and experience (and perhaps considerable guesswork), you allot 20 pounds to subsystem 1, 30 pounds to subsystem 2, and 50 pounds to subsystem 3. These preliminary allotments then become the objectives for these subsystems. Other allotments are also necessary. Consider reliability, for example. First, an allotment must be made to the subsystem, then to the components of the subsystems—all set up so that the overall objectives will be met. In these examples, the overall requirement is allotted (divided among) the several constituent parts of the whole.

In the design of a system, the *combination of input information* is by far the most important process. In fact, it plays a major role in almost all of our everday thinking.

Fig. 4-10 represents the combination of one or more items of information using some form of control information (actually, another input to the processing block).

In modern communications systems, many telephone, telegraph, and

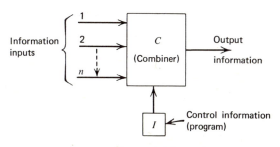

Figure 4-10.

television signals are transmitted simultaneously over a single transmission medium. The several signals are placed in proper relationship by wired-in filters or other arrangements that act as the control information that permits the "multiplexing." Multiplexing is very important in communication systems. It is of far less importance to you in system design.

In many of your activities, you take an item of information from here, another item from there, and a third from somewhere else. You put them together in accordance with some pattern. And you come out with a new arrangement. This is one way in which you can invent: you take some known parts or facts and put them together in a new way to perform an old function—or even to perform a new function. Note again, the two kinds of information that are necessary: input items (facts, ideas, and so on) and control information to determine the pattern in which they are combined. Without the pattern, you achieve nothing, even if you have the information. With the pattern, and without the input items, again you have nothing. Both must be present.

Another important example of combining is the modification, change, or alteration of an input. This type of processing can take many forms. For example, you may edit a letter and rephrase parts of it. The act of striking out unnecessary words (erasure) was described above under division. Here, we are talking about changing word and phrase structure, perhaps even changing the meaning. In editing, the control information is applied by your brain.

Now consider another step: language translation. You may take a document written in German and put it into idiomatic English. In this case, the input stream is the German version and the control information to convert it into English is supplied by your brain. Again, both kinds of input are necessary to provide the output.

The code clerk also performs this type of processing when he converts a message in English into code or cipher. He uses his code book to control the combination. And, of course, he performs the inverse operation to recover the original message when it is received.

Another alteration process is rearrangement or reordering of information. You frequently have to do this in writing. You move letters around, or words, sentences, paragraphs—even pages and whole sections. Again, your brain usually supplies the control information. The items of a drawing or table are frequently rearranged while it is being prepared and examined. Also, the pages of a book are collected from appropriate stacks and put together in the proper order before the book is bound. Thus, you can see that the word "rearrange" covers many different processes. However, all involve combination of information.

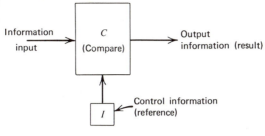

Figure 4-11.

Computation also involves combination of information. In performing arithmetic, two input numbers are added in accordance with a set of rules (control information). This is also true for subtraction, multiplication, division, and combinations of these operations. Also included under the general heading of computation are the logical operations that we perform almost automatically in our brains or that computers can be programmed to perform for us.

The final, important example of combining information is comparison. In comparison, an item of information is compared with a reference. In Fig. 4-11 the reference is the control information. Comparison with a reference value is the essence of evaluation. You use it to evaluate ideas, patterns, and proposals. You use it to check the results of all kinds of information processing. In the factory, comparison is used to evaluate objects, subassemblies, and assemblies by comparing their properties or performance with gages or special testing instruments. Management uses comparison to evaluate present performance and detect differences between what actually exists and what was predicted. In fact, evaluation is the basis on which you make the decisions that lead to your actions.

Decision and Action

After any evaluation, you are faced with making a decision. In making a decision, you are faced with three possibilities:

1. Evaluation is favorable. Positive action is indicated.

2. The evaluation is generally favorable, with some drawbacks or weaknesses. One possible action: a request for more information. Another: a request for a revision of the proposal. Possibly there is a way to modify or eliminate a cause (or potential cause) of trouble. Or

you may choose to plan some contingency or backup alternative—even with a warning signal of trouble. In critical systems situations, duplicate power plants are often furnished so that the system will not fail if one becomes inoperative.

3. The evaluation is unfavorable. Furthermore, it is so unfavorable that any effort at modification or revision is unlikely to be worthwhile. Under these conditions, the indicated action is to drop the subject.

Fig. 4-12 represents the basic block diagram for action after an evaluation. You receive the results of the evaluation in some form—by an observation, a written report, or by someone telling you. You process this information (think about it). Then you take some action. This action always involves control of your muscles to change some external object.

One particular kind of action is the direct control or adjustment of some instrument or system. You make such changes all the time. You pull down the window shade or tilt the venetian blinds when the light is too bright. You close a window to shut out a draft. You change your television set when you don't like the program. You steer your car to avoid the truck in front of you. In these examples, you perform a measurement or test, and make a correction. This sort of evaluation and action involves a "feedback loop" about which so much has been written. But the action output to control or adjust something need not be the result of your own observation. The information may come from an evaluation by something or someone else.

For example, you may set a clock to turn on your radio at a particular hour in the morning. The clock measures the passage of time and at the right moment closes the control switch for the radio. Thus, given a reference value (the time you want to get up), the clock makes a continuous comparison of the actual time versus the reference—when they

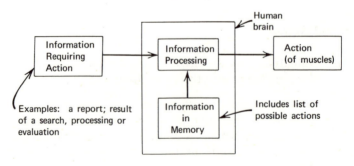

Figure 4-12.

coincide, it takes action. In other words, all such control actions do not involve a feedback loop, although many do.

Many decisions require action by other people. In these cases the information resulting from your thinking is either stored in some form or communicated.

Thus the end results of the gathering of information is an information output. It takes the form of a particular action by an individual or a particular communication from that individual which is either stored or communicated to others.

Summary

This chapter discusses the basic sequence of information processing in any design work. It makes much more precise such ideas as "getting the facts," "thinking things over," and "evaluating the situation."

An important concept is that the sequence is well-ordered in the mathematical sense: there is a first step that precedes all others. Thus, you *must acquire items of information* (facts) *before you can think about them. You must think about them*—no matter how briefly—*before you act on them.*

Remember that the order is not reversible. You can't think about facts until *after* you acquire them.

To store and retrieve information, a rather imposing list of requirements must *all* be met. The reasons why are pointed out.

Numerous functional block diagrams are shown and discussed. Furthermore, *all* of the possible kinds of processing of items input information are pointed out. For evaluation and decision, all of the possible outcomes are spelled out.

5

A System Design Procedure

The Information Pattern of a Design

In many ways, this chapter is the keystone of the book. First, it makes an extremely important point: in the course of a system design (or of any innovation), the amount of available information grows and grows and grows. Second, there is a pattern to the growth: to the acquisition and processing of the information. Recognition of this pattern leads to the "design procedure" discussed at length in this chapter.

Fig. 5-1 puts the information pattern of a design in a simple graphic form. When you conceive the initial idea, it can be explained very simply. Really, not much information is available. But with each successive step in the design procedure, more information becomes available. Finally, an enormous amount is necessary to tell how to build and operate a complex new system.

Now, it certainly costs money to acquire and process the needed information. A somewhat similar concept occurs in economics and manufacture: "added value." Initially, the materials of which the final product is made are truly "raw"—whether they are the products of mine, forest, farm, or sea. As they are processed, it is said that "value is added" by the work done.

The final results of a system design is a set of detailed and explicit papers. To prepare the set, information gathered from many sources must be combined with the small amount initially available. Some of the acquired information is discarded, some finds a place in the final design, and some must be modified before being used. Some of the end results may be drawings and specifications; others, the notes and conference reports telling of decisions made and the reasons for them.

To understand the fundamentals of system design, you must understand how to build up the information from a meager amount to the enormous quantity. This is the problem that we shall now discuss.

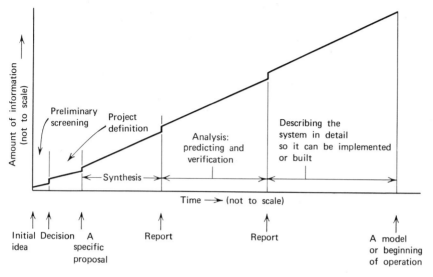

Figure 5-1.

Growth of Amount of Information in Design

Fig. 5-1 shows the growth of information as the design work progresses. In the beginning, when the initial idea takes form, little information is available. But as the work goes on, more and more information is acquired. When the design of a complex system is completed, tens of thousands of drawings and hundreds of pages of instructions may be necessary to contain the information. Even so, the files of letters and other documents may require an enormous amount of space.

As Fig. 5-2 makes clear, design is a step-by-step process. It is like a chain. No link can be broken. The design cannot succeed if any mishap occurs. There are many steps. Thus there are many possibilities for mishaps—or for inability to get necessary information. Hence, inevitably many promising ideas are never brought to completion. Some die in the beginning; some get a little way; some go much too far before they are dropped. Often insurmountable difficulties appear.

There are good reasons for the high casualty rate of seemingly good ideas. In the early stages, information may be scant. Often all that can be made available as references in making evaluations are estimates. Too often, these estimates may be purely subjective "hunches" or "horseback guesses." Later the facts come out, and the earlier optimism must be revised. Too often, the "good idea" no longer looks so good.

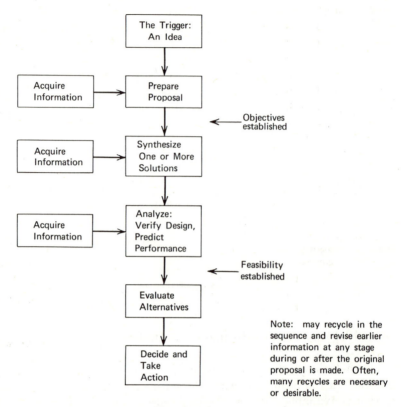

Figure 5-2.

Statement of The Problem

Almost always, a system design problem starts as a hazy idea. The functions to be performed (what is to be done) and the limitations on the design are stated in only very general terms—if at all.

In simple words, a system designer must do the following things:

1. Find all the necessary functions to be performed.

2. Find at least one apparently workable way to perform each and every necessary function. Preferably, you should find more than one way. If you can, then you can choose the best way.

3. Prove that you are right and that your solution meets all of the objectives.

The apparent simplicity of the three statements is deceiving. What you really need is a design procedure. As you will see, a design procedure actually requires quite a few steps. For some particular designs, some steps may be skipped. However, for a complex system, all are absolutely necessary.

Some authors have described what we are calling the system design problem as "concept formulation." No matter what it is called, you start out with a need or an idea. You define the problem and work out a tentative solution. You critically analyze each proposal. Weaknesses always show up. You revise the proposal to get rid of them. In fact, you revise, revise, and revise.

In this chapter, we shall describe a set of steps intended to solve a system problem. The basic steps are taken sequentially in time. The reasons will be pointed out. Furthermore, we shall discuss what each step is intended to do in advancing toward the final solution.

Why a Sequence of Steps?

As each step in the procedure is completed, more information is available. You critically analyze the situation as you get more information. Quite often, the additional information indicates desirable changes or weaknesses.

What do you do?

You go over the work already done. You make changes and revisions to improve the design or to remove the weaknesses—or both. As has already been said, always, in design, you revise, revise, and revise. Thus, each step in the procedure receives repeated attention.

Nevertheless, only when adequate and satisfactory information is available can the next step in the sequence be taken. Viewed in this light, the design procedure is an orderly method of obtaining and processing information.

The order of steps is fixed for logical reasons. The steps are not interchangeable. As you will see, this important fact can be quite useful when you learn to use the procedure.

In carrying out a design, normally you will go through the following basic sequence of steps:

Getting the idea.
Crystallizing a proposal.
Synthesizing a solution.
Analyzing and evaluating tentative solutions.
Comparing alternative solutions and selecting the best.
Reporting the results.
Building a model (if necessary).

Several of the steps can be broken into substeps. And, as explained above, during or after each step it may be necessary to revise earlier information. This revision may make it necessary to repeat one or more steps. Such recycling can be expensive in time and effort! The number and importance of revisions depends on the project and the difficulties encountered. Hence, the pattern of recycling can seldom be predicted.

Later in this chapter, we shall examine each basic step in order, and in some detail.

Some Time Factors

The time from the start to the finish of a system design may range from a few weeks to several years. For example, a minor change in an existing design may be relatively simple. A "face-lifting" may mean only small changes. Risk of failure is small. Consequently, the effort to get out a new design can also be modest. However, it may be difficult to add much engineering or design value to an old, competitive product. On the other hand, a successful, innovative design may offer an opportunity to contribute "added value" and reap resulting rewards.

However, if the new system is large and complex and involves many extensions of the present art, the situation always has uncertainties. For instance, a new supersonic transport requires new (and possibly untried) aerodynamic designs. It involves new materials (such as titanium). Sonic boom problems may be extremely difficult to solve.

The proponents of a new development are almost always optimists. Quite often, they have to be—otherwise the development would never be started. But even the proponents of the supersonic transport expect that the development will take years. Based on some past experiences, it may take more time than they expect.

The Trigger: An Idea

A new system design always starts with a real or imagined need. Remember the definition of a system says that it must perform some *wanted* operation(s). A new need usually requires a new system. Thus a want is created.

To be worthwhile, a new system must solve or simplify a problem existing in someone's mind. The ultimate use or benefit must be constantly kept in mind. Otherwise, a new system does not make much sense. From the management standpoint, new products can be a major factor in the growth of the company. They also affect profits. And, finally, new products are key factors in planning the future of the business. Growth industries are heavily dependent on the development of new products.

A successful new product has a characteristic pattern of sales volume and profit margin. Right after the product is introduced, the sales curve rises and so does the profit curve. But profits may tend to fall while the sales are still going up. Why? Because in many businesses, the older products are subject to cut-throat competition. In fact, sooner or later every product risks either being supplanted by something new or getting into profit-less price competition. For this reason, careful new product planning is essential to maintain profit margins.

Granted that new products are a necessary fact of economic life, much effort must be devoted to evolving one. Effort and care is necessary because it may take just as much time and money to develop a failure as a winner.

Once the need for a new system has been identified, an analysis should determine whether it merits further study. When management selects and develops a system to meet a new need, it is taking a major step. The risks of failure are likely to be high. The losses may be staggering. In choosing to meet a new need, it is choosing a number and kind of customers who have the need. It must take the competition into account. It must choose the suppliers and kinds of raw materials to be used in fabricating the new system. And, finally, it must make commitments of its financial, human, and material resources.

In this context a new system means a system new to the company. It may have been made in some form by others. But if the system is new to the company, it inevitably will raise new problems. The fact that some may have been solved by other companies does not change the situation.

Even a preliminary analysis may show that the new system is very different from those with which the company has experience. Maybe the possible returns are so high that the risks are justified. Maybe not. Only an analysis of all the facts can throw light on this question.

In the rest of this chapter a continuing example is referred to at the appropriate steps of the procedure. This example is intended to help you follow and understand the steps. The discussion will deliberately be made simple—even naïve. Otherwise, it would be easy to bury the real message in a maze of unnecessary details. Thus the discussion of the example *is not* intended to be complete—far from it! It *is* intended to be suggestive of an attitude and a way of thinking at each stage.

Let's imagine that you have a great idea: to build a new house for your family!

You need more room—in fact, you must have it. So the new system (the house) would fill a very real want.

Furthermore, as one simplification, let's assume you already have a lot on which to build your dream house.

Crystallizing the Proposal: The Initial Study

Most new systems (and most new products) begin with a rather hazy idea of a possible need and also some way of fulfilling it. In many cases, this is as far as they ever get. Infant mortality of new ideas is high. You can see why. Only worthwhile ideas should be followed through. The less spent on the others, the better.

To begin with, the hazy idea must be made explicit. Questions such as these must be answered: Exactly what do you propose to do? How do you propose to do it? Who will do it? How much will it cost?

Important objectives

Therefore, an initial study is required to set up some objectives. Among the important objectives are the following:

1. The functions to be performed by the new system.
2. *The performance requirements.* These may be conveniently divided into MUSTS and WANTS. In many cases, it is well to rank the WANTS in order of importance.
3. *System inputs and outputs.* For a preliminary analysis, you will need to know the wanted inputs and outputs. This is obvious. Not quite so obvious is the fact that you should think about unwanted inputs and unwanted outputs. Many a promising idea has been abandoned because of unwanted effects. The expression "unwanted side effects" is often encountered in the discussion of proposed new miracle drugs.
4. *The physical form of the new system.* Aside from the obvious factors of proposed size and weight, appearance may be extremely important. In fact, in architectural design, appearance is a major consideration. Considerable attention is devoted to the designs of the telephone instruments in your home and the equipment in your office.
5. *Cost.* With relatively few exceptions, a new system must make money for the company. For this reason, the answer to this question

is extremely important: How much will it cost? Of almost equal importance, in most cases, is the question: How much can we sell it for?

Constraints

When the study is authorized, you may be placed under many constraints that severely limit the design freedom. Some of the more commonly encountered constraints are the following:

1. Clearly, the inputs and outputs are important in the design of the new system. In fact, they set more or less rigid constraints on what can be done. It may be helpful to the progress of the study to arrange each kind of input and output in some order. For example, the most wanted could head one list; the most important of the unwanted, another list.

2. You may be told that some of the building blocks of the new design must be regarded as already specified. Sometimes almost all their details have been worked out before new design work is authorized. In an extreme case, you may be making a rather modest addition to an existing system: a system you must not change under any circumstances. For example, it may be in production. Only one or a few functions are to be added. Such a constraint may be frustrating because the job would be much easier if you did not have to work with the existing system.

3. You may be told what type of energy you may use in your design work. Some organizations prefer to use electrical components; others, hydraulic or mechanical. The reason for the constraint may be inherent in the nature of your system. However, it may result from company policies or your boss's desires. In either case it is still one of the rules of the game, as far as you are concerned.

4. Compatibility of subsystems may present important constraints. For instance, if you are working on a subsystem of a space capsule, your subsystem of the space capsule must work with all of the others. The other subsystems may be designed by companies a thousand miles away. Yet your design must work with theirs. For another example, the Bell System is designing new central offices. These new central offices must work with every existing central office. Some existing central offices are 40 or 50 years old. Yet the new office must be compatible with all of those in existence. As further examples, a new rifle may have to use ammunition of a certain caliber; a new weapons system must be installed in existing naval vessels.

People often talk about the "interface" between the two systems: for example, between the new weapons system and the naval vessel.

Any mismatch here will require modification of one or the other or some kind of matching arrangement. A new and larger torpedo for a submarine may require a larger launching tube. To install a new launching tube, alterations must be made in the hull. Alterations in the hull may require other changes in the vessel. And so on. All experienced designers have seen some of the unhappy results of a poor analysis of the compatibility of the new and the old.

5. You may be constrained in your choice of materials with which you can work out our idea. The slogan "Ideas take shape in aluminum" has already been cited. In particular circumstances, safety or cost considerations may govern the choice. Sparking contacts cannot be tolerated in a coal mine where they may ignite gases and cause an explosion. Except under special conditions, expensive materials such as gold and diamonds cannot be used.

6. Management may insist that certain devices or classes of devices be used (or not used) in working out the design. For many years, vacuum tubes were used for the amplifiers in all of the undersea cables designed by the Bell System. Why? Reliability. Tubes were used for a long time after transistors became available. Experience was available on tube performance over long periods. Recovery and repair of a cable laying several miles below the surface of the ocean is very expensive. Hence, transistors were not used until they had proved themselves. However, reliability is not the only reason for ruling out some devices. For example, it may be company policy to do (or not to do) necessary functions in a particular way. The policy may say that it will only use components manufactured by itself. If a new system requires components that it does not now manufacture, either the policy must be changed or the system cannot be designed.

7. Patents held by people outside the company are sometimes a constraint. You must either design around the patents, or the company must obtain rights, or the new system cannot be built.

8. Other constraints may involve people. For example, the manufacture of the new system may require the recruiting of many more people than the company has ever employed. Or, it may require people with quite different skills than those now employed. In either case, constraints may be imposed upon the design.

9. A final kind of constraint is the facilities that might be used for building the new system. It may be very important that certain particular existing facilities be used for the production. Perhaps they are nearing the end of a production run and need the work.

Management may want to use certain kinds of tools that they have on hand or that they can get at a good price. Or, based on the factory's experience, they may prefer to stamp out parts rather than to cast them. Naturally, a new design that requires additional plant construction or different tools starts under a handicap.

"Design for manufacture" is an important and often-neglected part of any system design.

Tolerances

In the preliminary design study, you must set up tolerances. Initially, actual numbers may be difficult to get. However, if the design is to be authorized, they will have to be forthcoming.

Just as the constraints are (or should be) set up, tolerances may be divided into MUSTS and WANTS; also, MUST NOTS and DON'T WANTS.

Tolerances must be set on the system inputs, outputs, and performance. What variations can be lived with? What variations cannot be lived with? Actual numbers must be guessed at or obtained.

The environmental conditions under which the system must operate must be set up. Typical requirements cover the temperature range, relative humidities, barometric pressures, and the tolerable variations in performance over the ranges.

You may be told that if your system costs more than $10 there will be no market. The $10 is a MUST requirement. At the same time, you may be told that the system *should* weigh less than 10 pounds. This is not a hard and fast limit. It is a WANT tolerance. WANT tolerances often imply an acceptable upper limit but do not say it in so many words. If the WANT requirement is 10 pounds and a design weighs a 100 pounds, it probably will not be acceptable.

The classification of tolerances as MUSTS and WANTS often permits design trade-offs. In the course of putting the design together and studying it, you may find that you can make it a bit cheaper if you add a little weight.

To return to your new home you need to set up the following important objectives:

Among the objectives is the number of rooms and baths:

MUSTS	WANTS
Living Room	Dining Room
Kitchen	
Bedrooms: 2	3 or 4
Baths: 1	1-3/4 or 2
Carport	Garage
	Family or Recreation Room
	Utility Room

Now for physical form:

> One floor?   Two story?   Split level?
> Basement?   Attic?

Appearance:

> Colonial?   Modern?   Spanish?   Tudor?

And certainly cost—a MUST:

> Not over $25,000 because of present and nearby
> future prospects of income and mortgage costs.

The constraints and tolerances are as follows:

Is electricity available?   Gas?   Sewer connection?
What do building codes say about:

> Size of home?   Location on lot?
> Materials for structure, roof, plumbing, etc.?

What about the climate?   Maximum and minimum temperatures?
Maximum wind velocities?   Snowfall?

Proposed approach to meeting the needs

You now need at least some information about how the requirements could be met. At this stage of the development of the idea, the information is very likely to be sketchy. Nevertheless, it must meet several requirements before you are safe in going any further.

Enough work must be done to throw light upon:

1. *Any problem areas.* Usually, problems result from lack of knowledge of a physical or chemical fact. For example, the design might require a new chemical reaction. It might require a mechanical subsystem of a type that has never been built. If the design is to go forward, then such gaps in the knowledge must be filled. Until they are, only guesses or estimates are possible.

2. *Evaluation of the risks incurred by going ahead without complete knowledge.* This can be a touchy area. Even experts can differ in their estimates of the cost of gathering information; of the likelihood of a needed invention; and of the difficulty in fabrication of a part by an untried technique. However, some thought must be given to the risks involved. Otherwise, you are taking a blind step in the dark.

3. *Alternative courses of action to minimize difficulties.* A good approach tries to foresee possible difficulties and provide suitable courses to minimize their effects.

You know that houses can be built. Your problem is to get a more definite idea about what *you* can build. So you inquire about building costs in your area (acquire information). You are told that you should figure on about $15 per square foot. Since your MUST cost is $25,000 maximum, your house must not have a floor area of over 1670 square feet. Thus a design is feasible with this as an important MUST NOT constraint.

Where you stand after this step

At this stage in the design, the intent is to establish the plausibility that a successful effort can be carried through. All of the constraints and tolerances are taken into account. Furthermore, if at all possible, several alternative ways of meeting the objectives should be worked out.

When this part of the study has been completed, then you know what you want to do in considerable detail. In fact, you are well on your way to having a specificatoin for the development of a system.

Communicating the Results: Selling the Idea

After you have assembled enough information, then you must communicate your results. Possibly, it is clear that further effort would only be wasted. In this case, you may have an unpleasant situation—particularly if the idea came from some influential member of the organization. If so, your report may make you quite unpopular. But facts are facts, and must be faced.

Suppose your studies indicate that the idea has merit: it looks like it can be carried through.

Now comes the hard part: getting an authorization to go ahead. This may be difficult because top management almost always is forced to choose between competing proposals.

In making these choices the stakes are high. There must be no major mistakes. If a good idea is turned down, and a competitor takes off and runs with it, losses may be serious. On the other hand, if the decision is to go ahead with an idea that should be dropped, then time, effort, and money spent will be wasted. When the stakes are so high, caution is essential.

In reporting your results, keep two important facts in mind. First, you should remember that the "company" is everyone in it—not just "management." Designers are part of a whole structure. Of course, you are—or should be—interested in making the structure operate as effectively as possible in pursuing its objectives and goals. If you are not so motivated, then you should be part of some other structure. Second, since you are part of the structure, you do not sell—just to "sell." You don't try to induce anybody (such as any "higher-up" bosses) to do something against their better judgment. Neither do you try to make them adopt a course of action without considering all the major factors. Instead, your purpose is to get the right decision—and get it efficiently. To do this, you present all the important facts—pro and con. Often you can leave out the details. Besides the basic facts, you can give your interpretation of them. This kind of presentation has many advantages. Your facts can be checked for errors of omission and commission. Your arguments can be followed. Improvements can be suggested. Good decisions require considering all the pro's and con's and the reasons for choosing among alternatives.

Three questions and some answers

In making your explanation of the desirability of going ahead, you must explain why the new system is necessary. What new needs does it satisfy? Why is it different? Competent management always asks three questions in one form or another:

Why do it at all?
Why do it this way?
Why do it now?

Let's discuss them in order.

Why do it at all?   One or more of several answers may be possible:

1. The preferred alternative design of the new system will make the company more money than any other proposal.
2. The new system will meet some threat by a competitor. Or it will satisfy a need expressed by a government agency or an important customer.
3. The new system will improve the company's image or enhance its prestige. Depending on the management's attitude, this argument may be important. But cars *are* built for handicapped people. Special telephones *are* supplied to deaf or blind customers so that they can communicate much as normal people do.

Why do it this way?   There are at least three possible strong reasons:

1. Assuming that a need exists, this is the best of the possible alternative ways of meeting this need.
2. The proposed system will have the best cost/effectiveness ratio.

As one Secretary of Defense characterized a missile system, "It will give the most bang for a buck."

3. The proposed system fits in with the company's present and future plans. The estimate of the resources needed to develop it are within the company's capabilities; the design offers an attractive way to use these capabilities.

Why do it now? Again there are some strong reasons, if they are applicable:

1. We must act before our competitors can act. Or we must meet our customer's or our government's requests.

2. This will employ labor or capital that otherwise might go unused. Except for the largest systems, designers are not often in a position to use this argument in a telling manner.

These are the kinds of reasons that can be and have been used to convince top management that it should go ahead. Thus they can form a sort of checklist in planning the sales job for your new system.

But remember: if you don't believe in the system, say so. If you do believe in it, do the best selling job you can. Always remember: far more systems are proposed than are ever authorized. Failure to get a go-ahead may be a bitter disappointment. But often top management must or does make the decision to say, "No."

In our continuing example of building a new house, at this stage you are both the planner-designer and top management. You have to re-examine the project and decide to go ahead or drop it.

Revise if necessary

As just pointed out, almost certainly objections will be raised against your proposed way of building the system. If you have done your homework, you will be able to answer many of the objections. However, quite often a thorough discussion of a proposed system brings out weaknesses or suggests better approaches. If so, then the proposal should be revised. Actually, almost all proposals are revised several times before final approval.

Where you stand after this step

Assuming that a go-ahead is given, the problem can be stated as follows:

Given. A specification of all wanted functions, unwanted functions, constraints on the design, and tolerances to be used to determine whether a proposed design is acceptable and to arrange in order alternative designs from the most desirable to the least desirable.

Find. A set of functional blocks to perform *all* the wanted operations.

A tentative proposal has been made telling how an arrangement might be put together to meet the objectives. Also, you should have a preliminary idea of costs; quite often, an estimated time to do the necessary work.

This work is sometimes called a "study" or "preparing a proposal."

Synthesis

Block diagram of the system functions

In attacking a new system design problem, sometimes it is helpful to start by drawing a block diagram showing the system functions. The need for this kind of diagram varies with the type of system. In setting up a new business, such an organization chart may be the only formal representation of the functions to be performed. This representation gives a logical picture of the organization that can be studied, criticized, and revised. Examples of such organization charts have been given in Chapter 2.

In the preliminary versions, some subdivisions may be intentionally omitted. For example, in a business organization, the initial interest may be focused on the manufacturing branches. Supporting branches such as legal and financial units may be left out of the chart. Furthermore, the number of levels shown on an organization chart may vary, depending on its purpose. Top management may concentrate its attention on the overall picture; intermediate management, on a particular department. In fact, in large organizations, many charts are prepared.

If organization charts are to be used in the planning of a system, they should show *all* of the wanted functions. To be sure that all are included, much information-gathering and discussion may be necessary.

As an example of the possible ways to arrange a set of such organization charts, consider a university. The overall chart should show all of the colleges and such supporting functions as admissions, finances, and plant. On this chart, the engineering college might (or might not) be broken down into electrical, mechanical, civil, and so on.

On such organization charts, numbers may or may not be included. For example, the organization chart of a college or a university may (or may not) show the number of professors, assistant professors, associate professors and instructors engaged in the activities of a particular department. Sometimes names and titles may be given. And sometimes other personnel such as stenographers, typists, clerks and laboratory people are also shown.

In other words, the amount of information to be shown on a chart depends on the intended use. It may be skimpy—or even intentionally incomplete. Or it may be quite detailed and show both the numbers of people, their positions, and their names.

Because top management is quite familiar with such charts, they are sometimes prepared as a step in the design of a new system. In many cases, they can help explain the system so that management can easily understand the proposed functions.

Such charts have advantages in certain cases. However, if a new design is only a modest change from an existing system, then such a chart may be skipped altogether. This is just another example of the old saying: circumstances alter cases.

Analysis of block diagram of functions

Any such organization chart that is prepared should be subjected to at least two tests: (1) A check to make sure that every block shown is necessary. (2) A check to make sure that the set of blocks on the chart is complete and that it portrays all of the functions to be performed.

Functional block diagrams

Although organization charts have their uses, for a complex system they do not give enough information. For better understanding, the inputs and outputs of each functional block must be shown and also the interconnections between blocks. Functional block diagrams (sometimes called input-output diagrams) are intended to do so.

When viewed correctly, the synthesis design process involves a selection of a particular set of functional blocks from all the possible sets of such blocks. In other words, it involves the selection of a particular system from all possible systems. Synthesis is the search for a system that meets a particular specification. The specification divides up the universe of all possible systems into those that are acceptable and those that are not.

The blocks required for the synthesis of a new system are all found in the set of fundamental building blocks given in Chapter 2. You may be tempted to say that seldom is anything basically new to be learned in system design. Only the particular application is new. Any contribution you can make results from your ability to apply knowledge to the creation of solutions to *new* problems.

The set of *kinds* of fundamental functional building blocks is small. On the other hand, the number of possible combinations of blocks is extremely large.

Despite these comforting thoughts, in real life to dream up a new system is likely to be an extremely difficult—even frustrating—task. Almost always, it requires experience and knowledge. Possibly a new system may result because some key components have become available. A classic example of this is the work of Babbage and Lovelace on their computing engine. Almost all the ideas that form the basis of modern computers were known to this talented pair. They worked hard for many years. Yet they did not succeed in building a computer. Why? Because they did not have available the devices that were actually used in the first practical computers.

A good idea does not guarantee that a new and useful system will be built—or can be built. Creativity may be essential. Often, invention.

Yet new systems are proposed, and new systems *are designed.* So we shall outline a workable procedure.

A basic principle in design is to avoid unnecessary work. For this reason, the specification requirements should be compared with properties of available systems. To do so, a list or catalogue of what is available should be consulted.

For a really new system, nothing that is available meets the specification. Hence, a new system must be synthesized. A set of blocks that together would meet the requirements must be found.

To get started, a very useful first step is to prepare a block diagram of the overall system. This block diagram should show the known wanted inputs, outputs, and operations. Of course, everything is not known—otherwise, there would be no problem.

To this block diagram should be added all known unwanted inputs, outputs, and operations.

Remember that the diagram should also indicate the environment in which the new system must operate.

Fig. 5-3 shows such a diagram for the case where the inputs and the outputs are given and the processing blocks are to be found. It is necessary to find a functional block or blocks to perform each wanted processing operation.

In filling in the unknowns, remember that every input (whether energy or object) must appear at the output. It may be changed in form. But the statement is true for both wanted and unwanted inputs.

A functional block diagram showing the flows through the system from inputs to outputs is sometimes called a "throughput flow diagram." It shows what happens to each kind of wanted and unwanted input

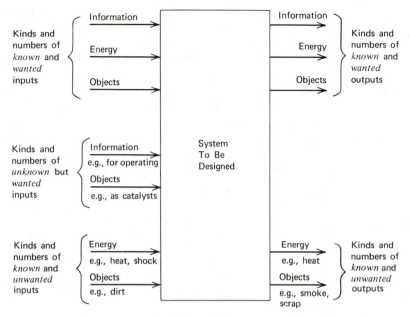

Figure 5-3.

as it goes through the system and appears as a wanted or unwanted output. Such diagrams are more or less routine in the study of power plants. Energy in the form of fuel is fed in. Part of the input energy appears at the output as electric power. Part is lost up the stack; part, in the condenser water output; in the case of coal, part is in the ashes.

You must make a complete list. And your accounting for inputs and outputs must also be complete and detailed.

Throughput flow diagrams are not as widely used as they might be. Properly applied, they can help to avoid errors that result from the omission of important design factors and by-products.

For some of the wanted processing, existing subsystems may be used or modified in a minor way. Others must be worked out starting with the elementary functional blocks and putting together new combinations.

In attacking a difficult synthesis problem, you may be comforted by the thought that a suitable functional block or combination of functional blocks can make any *possible* change in an input energy or object. The real question to be answered is: Is this operation possible?

As synthesis of a functional block diagram proceeds, control information flows are added. The necessary information sources from outside the system or from within (feedback) are shown.

Remember that if feedback is used, an unstable situation may be introduced. This possibility must be checked later in the analysis.

All of the necessary power sources for each of the processes are shown.

Finally, all inputs from and outputs to the environment are added.

As more and more information is gathered about a complex new system, a single overall block diagram becomes entirely too complicated to be useful. For this reason, you will break down the overall diagram into smaller diagrams. As the design proceeds, hundreds, thousands, or even tens of thousands of drawings may be necessary to describe the new system.

In preparing the functional block diagrams, remember that all functions must be accounted for; all necessary inputs and outputs and interconnections must be shown.

When it is completed, the diagram (or set of diagrams) shows all the arrangements of the functional blocks. Of course, as mentioned earlier, such a diagram shows all of the interconnections and interfaces between subsystems.

In synthesizing your dream house, basic functional building blocks are the rooms and other spaces such as halls (drawn accurately to scale). Together these blocks form the floor plans. Among the inputs and outputs of the several blocks are the doors and openings between adjacent blocks and to the outside and also the windows. In addition, the inputs and outputs for the electrical, gas, water, and sewage connections can be put on the floor plans at appropriate places.

Besides the floor plans, elevation sketches are also made to show the locations of doors, windows, and other important details that affect the external appearance.

Where you stand after this step

The diagram of the functional blocks that has now been drawn contains much more information than the earlier diagram of the functions.

The functions indicated are there because they are necessary for the solution of the design problem. Furthermore, they represent a proposed actual solution—not an imaginary solution. Of even more importance, they are not biased toward a particular, preconceived answer to the design problem.

It is desirable (and often possible) to prepare two or more alternative arrangements that can answer the design problem. Alternative proposals permit you to choose the best for the final answer—or even to take the best features of several alternatives, combine these, and come up with a proposal better than any of them.

For complex systems, this synthesis step is sometimes called "input-output" or "single-thread" design. All of the functions are shown only once on the diagrams. The numbers of blocks of each type required to produce the wanted *quantity* of output are not indicated. Such numbers are added later. Hence, the name "single-thread."

Some examples of other synthesis problems

You should understand that you will encounter other types of synthesis problems where the functional block diagram approach is not applicable. In such cases, you must devise an appropriate strategy for the particular case.

As an example, take an anagram. Suppose that you are given a scrambled set of ten English letters. This is your input. You are to rearrange them into a recognizable English word. The operation to be performed is the rearrangement of objects or signs in space. The output is the English word. The problem sounds simple. But the number of possible rearrangements of even ten letters is so large that unless you are lucky, it may take you hours or even days to solve the problem. It there are twenty or more letters in the anagram, then you may need a computer to help you out.

As an example of yet another kind of synthesis problem, take a mathematical proof. You are given the statement of the problem (the input). You are given the output: the QED. Furthermore, you are given the rules of the game: the acceptable mathematical or logical steps which may be used. The problem is to synthesize a sequence of steps so that you can get from problem statement to proof. This is the type of synthesis used by Euclid in his plane geometry. When you study plane geometry, you find his theorems worked out step by step. As you learn, you get problems in which you have to work out the steps, but the teacher knows that the problems can be solved. But, as you get older and more experienced, you find that sometimes neither you nor your teachers can come up with a suitable sequence. In mathematics, number theory contains many conjectures that seem innocent enough. Even a high school student can understand the problem. But they are only conjectures—and some have been conjectures for centuries—because no one has been able to prove them.

The two examples just given involve changing of information. But synthesis may involve the changing of objects. A designer may have to synthesize a machining procedure for making a new refrigerator, a new radio set, a new automobile, or a new missile. Or he may be asked to design a new process to turn a useless by-product of a manufacturing operation into a profitable product.

In the anagram and mathematical proofs, the problem is to find the right sequence of known operations. In changing objects, you may have to find both a set of operations and a sequence.

From functional block diagram to parts

At this time, functions to be performed by individual blocks or groups of blocks have been worked out. However, the functions are intended to apply to any particular way of building the system.

Next you must find out how to implement each of the alternative block diagrams with devices and parts that you can see and feel and operate.

The specification probably gives limits on the size, weight, and cost of the overall system. It is wise, before you start, to make a preliminary allotment of these overall figures to the several groups of blocks. Also, if you can, you should go further and put numbers for these factors on the individual functional blocks. Then, as parts for the blocks are tentatively selected, you can check to be sure that you are not going to exceed the specification limits for the system.

In this context, the word "method" involves two separate ideas. For one, the type of energy chosen to perform each function is very important. For example, some functions may be performed mechanically or hydraulically. For this reason, the possible types of energy for each function should be indicated.

The specification may set constraints upon the freedom of choice. It may specify that all operations be performed electrically or mechanically or in some other way. Clearly, if the specification says "electrical," then the devices in the new system must be electrical. On the other hand, there may be several possibilities. At this stage, all possible ways should be tabulated.

Even for a particular type of power, more than one method may be available to perform a function. Take a computer, for example. Suppose a computation requires the sine of an angle. The wanted sine may be computed by using an appropriate series. Alternatively, the computer may have a table of sines stored in its memory so that a particular sine can be "looked up." Again, all such alternative methods should be indicated. Furthermore, as new methods appear, they should be added.

The selection of parts for a system design involves a number of separate operations:

1. Choosing suitable parts or subassemblies to perform the wanted operations. These objects and subassemblies must have appropriate properties—mechanical, electrical, and so on. You do not use the same steel for a hair spring in a watch as for a bridge girder. The electrical

motors in cement mills must meet quite different requirements from those in office electric fans.

2. Choosing suitable stresses for materials. Consider the example of a steel beam used in a bridge. Are you going to assume a stress of 40,000, 50,000, or 60,000 pounds per square inch? As another example, consider the devices and subassemblies used in spacecraft. Materials can show some queer effects in the near-vacuum and the near-absolute zero temperatures of outer space. You must take these phenomena into account in choosing the stresses for parts exposed to such a cruel environment.

3. Considering how the design will operate under normal and anticipated abnormal conditions. This fact has been brought home to the automobile industry. Cars must not only operate efficiently and quietly while traveling 60 miles an hour. They must also incorporate safety features in case they have to stop abruptly. Steering columns must be able to absorb at least part of the shock in a collision. This is not an isolated example.

Every system designer must think about how his system will act when things are not exactly as he planned them. Bridges must not collapse under an unusual load; motors must not fly apart if the power suddenly changes.

4. In some designs, choosing or deriving formulas for computing such things as stresses and performance characteristics. Only when you have been up against many real-life design problems do you realize how many cases are not covered by the standard solutions you learned in your mathematics courses. When it is not in the books, you have to come up with the method yourself. Furthermore, after you have the method and want to use a computer, you may have to program the problem solution yourself.

5. In many design problems, calculating such things as:

   (a) the numbers of components or subassemblies required in the system;
   (b) the dimensions of components, of subassemblies, and of the system;
   (c) reliability of the system performance under normal and assumed abnormal conditions; and
   (d) costs of components, of subassemblies, and of the system. Since costs enter into almost every design decision, you must be prepared to make and defend your cost estimates.

6. Comparing and evaluating alternate designs. It is good practice to try to have more than one proposal that will meet the objectives.

Experience indicates that only in this way can the best overall compromise be selected. As part of your design effort you will be expected to prepare such comparisons and evaluations from the design standpoint to help management choose the best course of action.

Finding a suitable part or material to make a new part always involves information retrieval. If needed information is not available or can't be found, then the design is blocked until the lack is overcome.

In engineering design the usual source of device and material information is the lists and catalogs of suppliers. Some companies make many of their own devices; some purchase all or almost all of them.

If a necessary device or material is not available—or cannot be found—then the system design is in trouble. Only three alternatives are available:

1. Drop the design as unfeasible.
2. Discover, invent, or develop the needed device or material. The aerospace program furnishes many examples of parts and materials that did not exist until the need arose as the result of the program to explore outer space.
3. Change the objectives in the specification. Sometimes, such a change is a possibility. A WANT may be changed enough so that the design work can go ahead.

Experiments particularly directed to the finding of missing knowledge are sometimes undertaken. Such experiments are sometimes called *applied research*. Since the properties of a suitable part are known, an appropriate specification can be prepared. For example, many of the general requirements for the transistor were known long before the solid-state amplifier was invented and perfected.

Sometimes you may know of a suitable part but it is "unavailable." It may be made or used by a competitor, or it may be patented and your company has no right to use it. You may have to "design around" such unavailable parts. This may be possible by changing the method of performing the function so that available parts can be used.

In performing this step in the design procedure for your new house, you allot the total floor space you determined earlier to the individual rooms, halls, and other spaces. After settling on a floor plan, quite possibly you find that the estimated total floor space will not allow you to build as many rooms as you would like. So you have to revise your proposal—perhaps leave out a bedroom, or a utility room, or settle for a carport instead of an enclosed garage. Such cutbacks are quite common in building-type projects.

You choose the materials and dimensions for all of the subsystems of the house including, among other things:

The water supply, distribution and sewage (including outside hose connections).

The heating (and possibly cooling) system.

The electrical system, and so on.

Where you stand after this step

At this stage in the design and procedure, the feasibility of the proposed system has been established or reasons why it is technically or economically unfeasible have been disclosed.

Once a set of proposed parts has been chosen, then you can estimate the size, weight, and cost of the system. Preliminary work can be done on packaging and appearance.

The information developed in the course of the synthesis should give an idea about the difficulty of the development and of the time schedules. Furthermore, information should now be at hand to indicate more accurately the risks involved in going ahead.

Analysis and Evaluation

After one or more tentative designs have been synthesized and shown to be feasible, then each proposal must be carefully analyzed and evaluated. The purpose is:

1. To discover any latent weaknesses or faults in the design.
2. To determine the best proposal.

When you analyze a system, you must look for trouble spots. And you must be thorough. None must be missed. A searching analysis should disclose inoperative conditions

In a sense, you are trying to predict that the system will do what is wanted in every detail. And you know that prediction is a difficult task.

But the task also involves evaluation of competing proposals and optimization. In the present sense, the word "optimization" means securing the best performance according to some criteria. For example, when you tune your transistor radio to a wanted station, the best setting of the tuning knob gives the loudest response in the speaker.

Optimization is an extremely important part of designing. Almost always there is an optimum solution to a system problem; many or most of the characteristics of a system will have optimum values. Some of these optimum characteristics are stated as part of the system specification. You derive or decide others, as you proceed with the analysis and evaluation.

A word of caution. Many large systems are so complex that it is literally impossible to "optimize" them overall. There are just too many variables to handle—even if you could write the equations. In thinking of system optimization, you just don't find a minimum (or maximum) point by differentiating a simple low-order equation. You must try to determine some sort of unique "best" point in a multidimensional space of high-order. This may be very, very difficult! So you do the best you can—and be ready to defend your answer when you are questioned!

Other aspects of optimization are discussed later in this chapter.

Sometimes you find that two optimum values tend to conflict. For example, you may find that the lightest weight system costs the most money. In such cases, "trade-offs" are necessary. "Trade-offs" are part of the optimization procedure. We shall return to them later.

A check list

Analysis of this proposed system must give answers to a number of questions. The following questions occur so often in systems-analysis studies that they may be used as a check list:

1. Are all the functional blocks *necessary*?
2. Are *sufficient* blocks shown to perform all the WANTED functions?
3. Is the ordering of functions in space and time correct?
4. How many of each kind of block are required to produce the wanted system capacity under normal conditions? For example, in a machine shop, how many lathes, planers, shapers, and screw machines are needed?

For most systems, computations will be necessary to get these numbers. Quite often, assumptions about the work cycles of machines (times to perform each setup and processing operation) must be made. These may or may not be good estimates.

5. How will the system perform with multiple WANTED inputs in sequence, simultaneously, overlapping?

Again computations may be necessary.

The numbers of certain kinds of blocks may have to be adjusted so that the performance will be satisfactory.

6. What about the effects of UNWANTED inputs on the performance, including those from the environment? Will the system work under the extreme conditions required by the specification?

7. What about UNWANTED outputs? Are there any unwanted effects such as excessive noise, smoke, heat, or radiation?

8. How will the system behave under overloads? For example, how does your favorite restaurant operate when more customers want dinner than can be seated? Is there a silk rope? Or are they directed to the cocktail lounge? Or are they just told to go home? Traffic congestion problems (overloads) can be very complex in the design of toll booths for a tunnel, for a telephone switching system, or for an airline reservation system. Many other examples can occur in system designs.

9. How will the new system stand up under competitive conditions? Is it too vulnerable to enemy attack or jamming?

10. How sensitive is the system performance to variations of parts? This is a very pressing question in many areas of electronic design. For example, high speed digital computers may contain tens of thousands of parts. Naturally, because of manufacturing processes, all of the components of a particular kind are not exactly alike. So this question may be rephrased: Can the system stand the variations that are likely to occur in normal manufacture?

11. In what ways can the proposed system fail? And what are the resulting consequences? At one extreme, a leak in an air-tight space capsule will certainly cause the death of the astronauts. On the other hand, the failure of a part that is not essential for the system operation may be of minor importance. As an example, consider the failure of the light on the dial of your radio set. The radio still works.

Thus a complex system may have many failure modes and the consequences may vary from trivial to catastrophic.

12. What is the expected performance with partial failures of function?

A partial failure may result from a malfunction or a nonfunction. Typical causes include:

Wearout.
A premature random failure.
Marginal operation (e.g., low power voltage on tubes or transistors).
External operation (e.g., a blow).

To remove any weaknesses disclosed by study of items 11 and 12, the design plan requires some revision.

13. Is the proposed system compatible with any other systems or subsystems with which it must work? If you are designing a trailer-truck, will it be able to travel over the roads without violating any legal re-

quirements and meet any size limitations due to tunnels or overpasses on its route? If it is a new high-speed computer, will it accept the programs of the computer it replaces? Or do you have to design a whole new set of programs? If your proposed system involves actions by a human operator, is it compatible with his capabilities? To mention just two examples: Can he read and reach the knobs and dials that he must turn? Can he do it fast enough?

14. What are the anticipated costs of development, construction, operation, and maintenance? Costs are always important. Almost always, target values are given in the specification. Now that a preliminary system design is available, it must be checked to see how it meets the target.

15. How long will it take to develop and construct? Some excellent system proposals must be sidetracked simply because they cannot be made available in time to meet the anticipated market.

The list is not exhaustive, but it is a good start. You, as an analyst, must constantly think about other possibilities. A latent weakness that slips by and gets into manufacture can be disastrous.

In the literature, items 4 and 5 are sometimes called "high-traffic" design. This is an appropriate term for telegraph, telephone, highway, and airport systems. It is not appropriate for the other systems such as factories. However, the same *kind* of study is necessary for many systems.

If either item 6 or 7 reveals a potential trouble, you should change or add functional blocks to remedy the condition.

The examination called for by item 9 is sometimes called "competitive" design. The answers to such questions may be extremely important for military systems. For many others, they are unimportant.

Optimizing and trade-offs

During your analysis, you almost certainly will find some areas where the proposed design can be improved. Also, trade-offs are almost always possible and often absolutely necessary. This necessity can be shown by an example. Take the design of a new airplane. The specification says that it must be safe, fast, and comfortable. It must have a large passenger and cargo capacity. The first cost must be low; operation must be economical; and maintenance must be easy and infrequent.

Now if all your attention is placed on safety, other requirements will probably not be met. Why? Because if safety is overemphasized, speed and comfort must be sacrificed. Again, if emphasis is placed on speed, safety factors may be reduced, and so may the comfort of the passengers. High speed may require noisier engines and cause more vibration.

Thus you must strike balances between factors.

Striking a balance may be far from easy. How do you measure comfort? How much is noise reduction worth? How much is an added safety feature worth? How do you balance a lower speed against greater safety? It is clear that some trade-offs may be difficult to make because some factors are difficult to measure.

If you can find a common denominator such as dollars, you may be able to simplify your problem. But how do you measure added comfort in dollars?

Safety is a particularly difficult factor because it can involve trying to put a dollar value on a human life or lives (for example, in the case of a large passenger plane). But consciously or unconsciously, in deciding upon such trade-offs to optimize the design, you must have some feeling for the value of a human life. Otherwise, you may be so cautious that the plane cannot make money or possibly cannot even be built. After all, there is some risk involved in any flying. But difficult or not, the conflicts must be resolved. Some trade-offs must be made.

Again returning to your dream house, you submit your proposal to a careful analysis. You ask these questions:

1. Are the indoor traffic patterns optimum? Are bedrooms away from noise from the living and recreation areas?

2. Are bathrooms located to minimize plumbing costs?

3. Is the external appearance pleasing? Are views from living areas the best that can be had?

4. Are the pipe sizes for water supply adequate?

5. Are ducts and registers for heating (and cooling) well located and correctly sized?

6. Are there enough electrical outlets? Are the wire sizes adequate for appliance loads?

Almost certainly, you will want to get the best overall job you can for your $25,000, in light of your own objectives. To do this, you will do some optimizing—make some trade-offs. You may juggle room sizes to get an improved floor plan. For flooring you can make choices between materials: for example, between vinyl, oak parquet, and carpeting. You may leave the basement unfinished and spend the money on more expensive lighting and plumbing fixtures. And these are only suggestions to make the point. There are many choices.

Revision

As a result of the analysis and the trade-offs, a revision of the original design or designs almost certainly will be necessary. These revisions may involve the materials to be used, the parts, the functions shown on the block diagram, or even the contents of the original specification. In fact, a careful analysis and optimization may result in changes in all of these. The result should be a better proposal or proposals for the new system.

Where you stand after this step

Hopefully, after so much effort has been expended, one—or, preferably, more than one—feasible proposal is available for the new system.

If a feasible proposal has not emerged, then there are four possible decisions:

1. Redirect the work so that it will meet the original specification. For example, set up a different functional block diagram. The basic objectives and program for meeting them were not at fault; the proposed implementation was at fault.

2. Change the specification.

3. Review the decision procedure. Possibly, a better course of action was available (perhaps a different type of activity for the company) but was not chosen.

4. Drop the project. This decision may be painful. But it may be necessary because the work has shown that the proposed system is not feasible, is uneconomical, or cannot be produced in time. All that has been gained by the effort is information and experience.

Compare and Select

If more than one proposal meets the specification, the best must be chosen. A choice can be made by using the results of the analysis and decision rules, the WANTS and DON'T WANTS, contained in the specification.

One possible action is often taken at this point. The best proposal is further revised to incorporate the good features of two or more superior alternatives. The final version is the result.

As pointed out earlier, the selection is often based on dollars: costs, dollars expected profit, or percent return on investment. If two or more important values are involved (such as dollars and performance), the selection may be more difficult. But, in any case it must be made.

This is the stage in the design procedure of your dreamhouse example at which you should make a basic decision:

Shall we go ahead and build?
Which of the alternative designs that we like shall we choose?
Should we recycle and try and get a still better design?

Hopefully, after all of the information gathering and processing that you have done, you can unhesitatingly decide to go ahead and build a model—in this case, your new home.

Where you stand after this step

At this stage, the best proposal has been selected or it is known that none is satisfactory. Unless it is necessary to build a preliminary model, the design activities are completed—except for the report.

Report on the Design Work and Principal Results

At this stage, almost always a written report is prepared. In fact, a good report is one of the most important products of the design effort. The comprehensive overall report on a complex system may be lengthy. Also, you may be called on to give one or several briefings to top management on various aspects of your work.

Whether you submit a written report or present the results orally, you must be prepared to convince your readers or listeners that you have offered the best possible answer. In fact, before you can hope to get an approval of your ideas, you must sell them. Management hears many proposals and must select only the best ideas. Hence, your audience may be skeptical. Therefore, you must be prepared. You must anticipate possible objections. And if you are wise, you will have thought through the answers to the objections that seem most likely to be made.

For a written report, you should include:

1. The system specification (possibly as an appendix) with only the most important requirements restated in the body.
2. The decision rules you have used in choosing the best proposal.
3. The most promising alternatives for the design. Perhaps the drawbacks of discarded proposals may be included.
4. The estimated performance of the promising proposal. Depending on the report, only the most important aspects may be covered, or all of the estimates may be spelled out in detail.

5. The reason for the selection of a particular proposal, or the reason for rejection of all proposals.

The final version may contain a summary of the actions decided upon and the reasons for the decisions. Often the final decision may not be taken until quite a while after the rest of the report has been written and revised.

If the proposed system involves patentable ideas, these must be brought to the attention of the patent lawyer. Quite often system designers generate patentable ideas.

Finally—and this depends on company policy in most cases—you may be asked to write one or more technical papers. Many large companies have their own publications in which descriptions of new systems are presented. On the other hand, the company (or a military service) may regard publication as unnecessary disclosure of "secrets."

A Model for Tests (if Necessary)

Remember that the design work has produced information defining the new system. It states the WANTED and UNWANTED inputs and outputs; the operation of the overall system; and any restrictions such as size and weight. A selected set of subsystems has been worked out together with their inputs, outputs, and WANTED operations. The design work has made it fairly certain that all of the wanted subsystems can be built.

Frequently, a model must be built. In this connection, we are not talking about a mathematical or simulation model. We are talking about actual hardware. The purpose is to gain assurance by tests on the model that the project is sound.

This model may take a number of different forms, depending on the wanted information. For example, it may be a scale model. A scale model of a new airplane can be tested in a wind tunnel. A scale model of a ship's hull can be tested in a model basin for its performance in various kinds of seas. Quite often a scale model is made of a proposed new building or group of buildings such as a shopping center. At the other extreme, a full-sized mock-up may be made of the fuselage of a new airplane. Then, different seating arrangements can be tried out and other important factors studied.

In still another type of model, all of the functions may be present so that they can be tested. However, if the full-size project might have, for instance, 100 machines of a particular kind, only about 6 would be provided for a "skeletonized" model. Obviously, fewer machines cost less. A pilot plant of a new chemical plant can provide facilities for checking all of the expected reactions of a full-size plant but can utilize much smaller quantities of chemicals in doing so. The smaller quantities mean smaller pipes, and smaller retorts and other reaction vessels.

Still another type of model may be a combination of items of critical hardware plus computer simulation of particular aspects of the operation.

One way to communicate the requirements for the model is a set of engineering drawings. These may be carefully prepared in great detail. In fact, they may be the chief means of documenting and communicating the results of the engineering design to the people who actually construct the model. They contain all of the dimensions of the parts to be made and spell out in detail how the assemblies of the subsystems and the system itself are to be carried out.

Once made, the set of drawings is turned over to the people who do the actual construction of the model. In some companies, the liaison between the designer and the shop people is quite close. If this is the happy situation, when troubles come up in the drawings (and there always are troubles) they are worked out in a talk between the designer and the shop people. In fact, with good liaison the drawings may be quite sketchy.

After the model has been constructed, it is ready to perform its primary function: to provide information. To get the information it must be tested. Furthermore, the results of the tests must be interpreted.

In large organizations, the tests are not under the control of the designer. In fact, they actually may be made by a different organization at a different location. For this reason, we shall not give a detailed description of how they are carried out.

In almost all instances, the test results reveal weaknesses or desirable changes to improve some aspect of the performance.

When a change is necessary the designer again comes into the picture. A weakness may require revision of the results of any previous design step. For example, the tests may indicate necessary changes in the specification, changes in the functions of the subsystems, of parts, or of materials.

When the tests indicate the need for the revision, the designer works out a new proposal—perhaps in consultation with the test people—and this is incorporated. Further tests follow to make sure that the change remedied the difficulty.

Revision may follow revision until, at last, the model is satisfactory.

Where you stand after this step

After the model is satisfactory, then and only then are you reasonably sure that the design proposal is sound and can be built. The design work is over. A new system has been designed.

Summary

The design of a new system (or of any innovation) is characterized by a growth of information from a small initial amount to a much greater (sometimes tremendously greater) amount. The cost of the increase is analogous to the "value added" in manufacture of a physical object—such as an automobile.

Conceivably, the growth of information might be an entirely random process. Actually, it is not. True, luck or serendipity occasionally play a part in an innovation. Nevertheless, there is a step-by-step procedure that should be followed to guide the design process.

The steps in the procedure are ordered. Why? Because you, as a designer, can only process the information available to you. Growth can only progress in one direction.

The first step in the ordered procedure is *the idea*. Obviously, without an idea of what you want to do, progress is impossible. Therefore, you always must start with an idea. Initially, the idea is just that—it lacks detail.

The next step is to develop *the necessary details*—to put flesh on the bones. Also, for the idea to be practical, there must be an insight into some way to implement it. The details of the idea and how to realize it may or may not be put in the form of *a written or oral proposal*. This step must follow getting the idea—it cannot precede.

Given a formal or informal proposal, *the next step is synthesis:* finding some combination of information, or things, or both that together promise to meet the proposal. Also, after putting together a possible combination, it is desirable to check it to see that all the elements are necessary and that they are sufficient to meet the requirements. Preferably, several alternative possibilities should be prepared.

With one or more tentative syntheses in hand, *the next step is analysis* of each one to uncover any conceivable weaknesses or shortcomings. Almost always, some necessary or desirable changes turn up.

Next is evaluation of each contender, *and selection* of the best. This step is only possible *after* synthesis and analysis.

At last, *the final results* of the design work *must be communicated* either orally or as a written report. Perhaps interim progress reports have been made upon occasion during the course of the design.

Depending on the circumstances, *a model may or may not be built* and tested after the design is complete or nearly complete.

At any time after the original idea is conceived, new information may indicate that changes should be made in the results of some preceding step: a recycling to the earlier step, and some rethinking. For example, a weakness disclosed by careful analysis may require changes in the synthesis or in the proposal, or in both.

Recycles are a characteristic of innovative work. They do not change the basic procedure of the design sequence: they mean that steps may be repeated—possibly several times.

Each step in the design procedure is discussed in its proper place in the overall sequence. After each step a recapitulation spells out what was accomplished.

A familiar continuing example is intended to help you follow the development of the exposition.

Several checklists are presented. They have helped others in their design thinking. They may help you too.

6

Some Other Design Aspects

Introduction

As a designer, you must acquire, store, and process the information that you need (or may need) to do your work. This is by no means a trivial chore.

Also, you have another important role: communication. You must be able to talk to other people about your ideas, your work, your difficulties, and your progress. You must be able to write reports.

In large organizations, top and intermediate management also make important contributions to a successful design. In the present context, the best dictionary definition of "to design" is "to make designs or plans" and "to adapt means to an end." Also, according to the dictionary, "to manage" is "to control or direct the movements or workings of; to direct or conduct the affairs of; to carry on business." Furthermore, management is defined as "the skillful use of means to accomplish a purpose."

The dictionary thus makes it quite clear that design and management are quite different activities.

In a successful arrangement, design and management work should complement each other. They should strive to achieve common objectives and goals.

This chapter discusses some of these aspects of a design endeavor.

Designer's Roles

Collecting and storing information

Your effectiveness as a designer depends, to a great extent, on how well you acquire, store, and retrieve information. You should set up your own file for the information you need frequently. Your files should also contain useful drawings, specifications, and other papers arranged so that you can refer to them quickly and easily. Preferably, you should set up some form of indexing system.

A good technical library can furnish you much helpful information. Technical books and the handbooks pertaining to your work are a *must*. Also, to keep up to date, you will want to have technical magazines. If your company has a large library, you should use it. But trips to the library can take time. And the book or magazine that you want always seems to be "out" just when you need it the most. To minimize these frustrations, you should have some books and magazines of your own.

As a designer, you are responsible for keeping up with the advances in your field. This may mean that you attend seminars or take college courses. A designer cannot let himself get behind the times. He must always keep up. And keeping up means some form of continuing education. The basic courses in mathematics and science that you learned in college are always a foundation on which you can build. But mathematics that you don't use every day may just silently slip away. And advances are made in the basic sciences from time to time. Hence, in one way or in another, you must refresh your memory and enrich your knowledge of the fundamentals so that you can use them when and if you need them.

Besides the knowledge and information you possess, the knowledge of other people can often be of enormous help. A friend who is interested in your work sometimes can supply a bit of missing information. With this information, everything falls into place: you have an answer. Even if friends are not experts in your field, talking to them and explaining what you are doing may enable you to understand your own work better. A friend who is a good "sounding board" is a good friend, indeed.

Along the same line, some large companies have consultants available for particular fields of knowledge. If the consultant is a mathematician, he may help you put your problem into a form so that the equations can be solved. Or if you have the problem in mathematical form, he may show you how to solve your equations, or solve them more easily.

Finally, as an active designer, you may find it desirable to join one or more professional societies in your field. If you are outside the large metropolitan areas, you may find it difficult to attend meetings, so that the major advantage is to receive the organization's publications. Unfortunately, many of the professional societies cover such a wide area that the amount of information you can gain is small compared to the flood of paper that you receive. Finding a useful idea in this flood of material is often like looking for a needle in a haystack. Besides, you are not always sure that there *is* a needle in the haystack.

As a designer, never forget that you must keep up to date—even in the most stable industry. You may make your career with one company. Even so, you must expect to be put into a new job every few years. The new job means that you will have to gather, store, and retrieve new information. But if you know how to do this, and how to process information, the new job will be challenging—but you can face it with assurance. During World War II, physicists designed the atom bomb. Mechanical engineers made it work. Communication engineers worked on radars, sonars, and acoustic torpedoes. These designers had no previous experience in these fields. They had to learn—and learn fast.

If you expect to change companies every few years, then obviously you must revamp your information. True, you may get your new job because you have some experience that your new employer needs and wants. But your past experience is only an entering wedge. It will not guarantee your future indefinitely.

To sum up, you must be adaptable in today's world. And when you make a change (or a change is made for you) you must be agile. If you are not, you will certainly lose out. You must keep your information up to date, and your processing skills must never be allowed to get rusty.

Communicating

As a designer you have another important role: communication. If you get a new idea, you must know how to explain it. Otherwise, you might just as well not have it. Selling is never easy. And selling your new idea may be the toughest form of selling. Yet it must be done.

If you want rewards and advancement, you must be prepared to explain and defend your designs. You must make your expositions crystal clear to your audience. You must show why it is the best possible design. If weaknesses are pointed out, you must acquiesce and revise your work. Perhaps you may have to explain why it took so long and cost so much to work out the proposal.

Then there are reports. You should make reports to yourself: reports of what you did and, even more important, why you did it. These may be merely notes—or they may be a daily log. In any case, they are essential. You can't depend on your memory—no matter how good it is. Details become fuzzy; important facts can be forgotten.

Particularly in a large organization, you will be making many reports to your colleagues and superiors: what you are doing, why you are doing it, and why you are not doing it some other way. Your superiors get their information about what is going on from you. And their method is by asking questions and listening to reports.

Yes, communication is an important role for you. Conferences, seminars, laboratory reports, progress reports—the list sometimes seems interminable. To gather your information, you must know how to acquire and retrieve information. To communicate what you know, you must talk clearly and effectively, and write like a professional.

Now, we are back to the need for self-education. If you are not well skilled in reading, writing, and speaking, then you need to improve yourself. Otherwise, you are handicapping yourself in playing all of the roles that you must fill—and fill well—if you are to be top flight.

Management's Roles and Responsibilities in Design

As a designer you may propose and synthesize a plan, analyze it to make sure that it will work the way it should and, possibly, even test a model. On the other hand, top management selects, sets up, and controls activities. A manager's information output is a decision. After a decision is made, an appropriate action (or no action) follows.

In fact, whether you are a member of top management, intermediate management, or a designer, your major activity is to process information. To process the information with which you deal, you must use your brain. You must think about it. And your thinking will follow almost the same patterns whether you are in top management, intermediate management, or design. The subject matter will be different. But the techniques you use in handling it have much in common. The points of view are different. In design, you think about performance, safety, reliability, and related areas. In management, you think about sales, profit, corporate progress, budgets, money and, above all, people. However, regardless of your level, you should try to understand and help those to whom you report, and those with whom you work.

In a one-man shop, or a small business, all—or at, least, almost all—of the roles and responsibilities are there. Few—or very few—of the functions depend on the size of the organization. Suppose that you are an artist. As your own manager, you must think about sales, progress, and money. Then you change your headgear, and think about forms, shapes, colors, and materials that will convey the images in your mind to the beholder of your painting.

The manager and the designer (each in his own way) deal with both the present and the future. Both work in uncertain worlds. Both must predict and take chances that their predictions are right. The designer may find that the performance he needs just cannot be achieved. And another brave new idea has to be put on the shelf. A manager finds that the new product just does not sell.   He took a chance and lost.   But sometimes the design works out.   Sometimes it makes a profit.   Of course, this is the hoped-for outcome.

Fortune magazine, in typical *Fortune*-like language, explains that management's guiding principle is that results are accomplished by committing resources to exploitable opportunities—not by solving problems. But during the design work, management must watch the progress.   It must measure the accomplishments and compare them with the plans.

From the first glimmer of the idea to the working model, management makes the decisions to go ahead or stop.   Management may decide to stop at any stage for any good reason.   The more work done on an ill-conceived proposal, the greater the potential loss of money and time.

Summary

As a designer you must acquire, store, and process information.   To do this effectively, you should set up and maintain your own files of material that you are likely to need.   But you must do more: keep yourself up to date by study and, possibly, by attending technical meetings.   You must constantly restock your mind with the latest findings in your field.

Besides being competent in your thinking, you must be able to communicate effectively: to write well and tell what you know convincingly.

As a designer, you must also be able to manage yourself in your work.   Remember "to manage" means "to control or direct."

If you are in intermediate or top management in a large organization, you must not only manage yourself but also those for whom you are responsible.   Although you are in "management," you still have to design.   You still have to process information.   The subject matter changes.   The points of view are different.   But your thinking follows much the same patterns, regardless of your level.

7

Techniques of Design

Design and Thinking

By definition, a system consists of a set of components arranged to perform some *wanted* operation. The word "wanted" is significant because it indicates that someone wants the operation performed.

In designing a new system, you must innovate. You may come up with a new combination of known components to perform a particular operation. For example, an electric shaver performs the operation previously done by an "old-fashioned" straight razor and, more recently, by "safety" razors. Alternatively, you may dream up some entirely new operation. You may find this hard to do because men have been dreaming up operations for centuries. Women washed dishes long before electric dishwashers. Many great new ideas are the creation of the advertisements; few are really new.

To a system designer, the word "wanted" implies another important concept: that the results of his work will be used. And use is the highest form of recognition he can desire. It is hard to derive pleasure by working on something nobody wants.

John W. Haefele, in the preface to his book *Creativity and Innovation,* defines creativity as "ability to make new combinations of social worth." Since the design of a new system requires innovation, the secret of success lies in the words "ability to make new combinations."

How do you arrive at new combinations? By thinking—and only by thinking. Rudolph Flesch, in his *The Art of Clear Thinking,* defines thinking in this way: "Thinking is the manipulation of memories." In other words, you innovate by putting together facts that you know (or imagine) in some new or unusual kind of pattern.

This concept is beautifully simple. The execution may be extremely difficult. This chapter gives suggestions and techniques appropriate for each stage in the design process: from idea to working model.

By definition a successful design effort necessarily means a successful outcome. Certainly, the outcome may not be exactly what was originally intended, but it is still both wanted and useful. Under these assumptions, we can define an efficient design procedure as one that requires:

1. Minimum effort and mental stress.
2. Minimum use of resources (money and facilities).
3. Minimum time for completion.

Hopefully, it will be very successful or, at least, it will have the best possible outcome.

Sequence of Operations in a System Design

Previously, the sequence of operations in system design has been discussed in some detail. A somewhat simpler presentation is sufficient for our discussion in this and the next chapter. Fig. 7-1, beginning at the top, shows the steps to be discussed. The parallel arrangement of the design effort and the management contribution is intentional. Repeated reference to this parallelism has been made previously. This is done for a very good reason: the two kinds of effort must go along simultaneously; each must support the other. Yet they are fundamentally different in their point of view.

In the literature the concept is sometimes expressed that system design really is a "tangle of problems." Furthermore, in the literature of management science, the allegation is made that many of the problems are poorly stated and poorly structured.

True or not, the problem that faces you, as a designer (or manager), is to bring order out of chaos. If you are to be successful, you cannot sit around and bewail the fact that the problem is not clear or not well worked out. You must get an idea—an idea that makes the situation clear!

Once you have the idea, then the sequence follows.

Designer and Management Roles

To repeat: both designers and managers collect information, both evaluate information, and both process information. Until a model is actually available for test, no hardware is involved.

You, as a designer, are responsible for the system performance. As

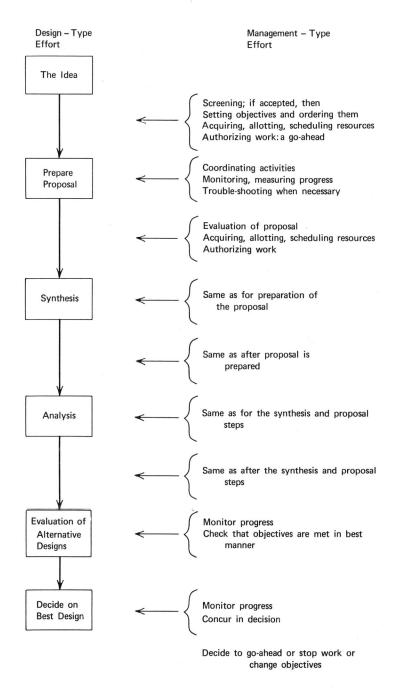

Design – Type
Effort

Management – Type
Effort

The Idea

Screening; if accepted, then
Setting objectives and ordering them
Acquiring, allotting, scheduling resources
Authorizing work: a go-ahead

Prepare
Proposal

Coordinating activities
Monitoring, measuring progress
Trouble-shooting when necessary

Evaluation of proposal
Acquiring, allotting, scheduling resources
Authorizing work

Synthesis

Same as for preparation of
the proposal

Same as after proposal is
prepared

Analysis

Same as for the synthesis and proposal
steps

Same as after the synthesis and proposal
steps

Evaluation of
Alternative
Designs

Monitor progress
Check that objectives are met in best
manner

Decide on
Best Design

Monitor progress
Concur in decision

Decide to go-ahead or stop work or
change objectives

Figure 7-1.

a designer, you must realize where you are in the sequence of steps in Fig. 7-1. To mention a few examples, synthesis is not analysis. Analysis is not searching for information. The tasks are entirely different. And, finally, searching is not selling. And you always have to sell, sell, sell: your ideas; your synthesis; and your analysis.

You know now that a management function is to make the decisions: for example, to go ahead with an idea or drop it. Management is always the evaluator. And the evaluations include designer performance.

In preparing your design, you must also make decisions. Thus you "manage" yourself and your work. The degree of your success depends to a considerable extent on how well you do this.

Large organizations have quite a few layers of "management." One or two levels above the designer you often find a "project engineer" or "project leader." He combines some management with some design roles. He may "design" strategies for attacking problems; he "designs" presentations; and so on. But he also evaluates performance, and he often decides between alternative proposals. Sometimes he refers vital decisions "up the line" to higher management.

Even the highest-level managers also combine the management and design roles. For example, they "design" the company organization; and they "design" marketing programs. But they certainly make all of the high-level decisions such as those involving the allocation of resources of men, money, and facilities.

Thus, at every level, you find that both design and management roles must be played. However, the subject matter to be processed differs from level to level.

Conversely, in a one man shop (such as a painter), all of the design and management duties must be performed by the one individual.

Both designers and managers deal with information. But, in many cases, they place different emphasis on particular items of information. Their viewpoints are different.

The requirements for designers and managers differ so much that large organizations often—even usually—have different people to perform the two functions. Large enterprises have found—sometimes by bitter experience—that few individuals are good enough to take over full responsibility. You must have all the necessary knowledge to do an outstanding design job. And management talent is also a scarce commodity. Even more, a good designer often has no desire to take on management responsibilities. One mistake you will observe in many large organizations is the advancement of a good designer to a management position which he does not want and which he does poorly. Often, a company trades a top-flight designer for a mediocre management person.

Importance of Facts and Relations

In Chapter 5 the required information for each step in Fig. 7-1 was discussed in considerable detail.   Also diagrams, made up of fundamental system building blocks, were shown to indicate the information that you must have before a successful outcome can result at each step.

The raw material on which designers and management work are facts and relations.

With your mind you process this raw material—you "think" about it.   You may think long and hard.   You may come up with an answer; or you may not.   You may have an "inspiration" that answers all the questions in one fell swoop.   In your thinking, you may call upon your imagination.   Emotions (for example, frustrations) may hinder you.   Also, emotions are sometimes aroused in designer-management situations.   A designer may want to get just a little better performance out of his system.   Management may say, "No."   A conflict is possible.

Divergent and Convergent Thinking

Some psychologists define two kinds of thinking: divergent thinking and convergent thinking.   In divergent thinking you try to generate a lot of ideas.   You do *not* try to solve a problem by finding the final answer.   Returning to Fig. 7-1, divergent thinking is the type required for the "idea" block.   This kind of thinking is particularly applicable when there are no clearly right or wrong answers.   Proposed answers can offer various degrees of objective and subjective adequacy as a possible solution to the problem.

Divergent thinking is clearly a way to tackle the problem of getting a "new" idea.   Advertising people use it to try and get a new slant on such old household products as soaps and detergents.   You as a designer—or you as a manager—also occasionally need this type of thinking in examining the possibilities for a new system or a new use for an existing system.

You should use *convergent thinking* when you must come up with *one* or only a few solutions to a problem.   You use it when you have a clearly defined goal and your efforts either succeed or fail, depending on whether you achieve that goal.

Convergent thinking is the tool that you use in synthesizing a new system.   Your efforts either end up with a suitable answer (or answers) or you fail in your objective.   In mathematics, you either find a proof for the theorem or you do not.   The results are clear cut: yes or no.

You also use convergent thinking in analysis and evaluation of a proposed system. Your goal is to say the proposal will or will not work. It can (or cannot) be made within the cost, size, weight, and other applicable constraints.

A manager also uses convergent thinking when he is confronted with a situation that needs correction. He must isolate the cause, make a decision, and take action.

Thus there is a basic difference between the two kinds of thinking. In system design, you must always know whether you need many ideas or a solution. You are in trouble if you don't know which you want. Furthermore, psychologists have found that some people never seem to be able to do convergent thinking. They can never concentrate on a problem long enough to finish it and get the answer.

The rest of this chapter discusses these two quite different types of thinking and the techniques that are applicable. From what has just been said, you can see that their areas of application differ. So do the techniques. You should understand both of them, and you should also understand when each should be used.

Both designers and managers use both types of thinking. Their use will first be discussed from the designer's point of view, and then from the manager's.

Getting an Idea

What is a good idea?

According to Rudolph Flesch, a good idea starts when you have a seemingly insoluble problem. Suddenly a neat, simple solution comes to you! You had a bright idea! And if your idea is really useful, you may be flattered by someone saying, "Why didn't I think of that?"

Yes, Flesch is right, ideas come from problems. But in the early stages of design work, the object is not always to find one final solution. Often a lot of ideas are wanted—at least initially.

What is needed is "divergent" thinking. Even far-out ideas are acceptable. There are no completely implicit or rigorous standards of acceptability. There are no clearly right or clearly wrong answers— merely various degrees of adequacy as possible solutions.

In system design, at first the functions to be performed may be quite hazy. The exact limitations may be unknown. Only the general features of what is to be accomplished are sketched out. Certainly when the idea of going to the moon was accepted by President Kennedy, he had no idea of all of the functions that had to be performed. He had no knowledge of the difficulties that would have to be overcome. But the objective of putting a man on the moon in the decade of the 1960's was a "good idea."

As he knew, the necessary details could be spelled out in specifications only after much careful thought.

There is no foolproof procedure for getting good new ideas. But if you understand the concepts set forth in this chapter and apply them, they can pay off for you. You should be able to get more good ideas and improve your batting average when confronted with the need to come up with them.

Sources of new ideas

Where do new ideas come from? From the human brain. And only from the human brain.

When you are confronted with the need for a new idea, you search your brain for relevant experiences. You fall back on your storage of facts and patterns. In some mysterious way, you change one or several facts, or you put them into a new and different pattern. How you do this, nobody knows.

But we do know one thing: association is a very important component in getting a new idea. In this connection, the word "association" is used in the sense in which psychologists give "word association" tests. You answer a test word with the first word that flashes into your mind. It is assumed that, in some way, these two words are "associated." For you, as an individual, your associations depend on the sum of all of your relative past experiences involving the test word. Your associations may be off-beat—not those that any other person would have. When the test word is given to you, it "triggers" your response.

To the extent that association is involved in getting good new ideas, off-beat associations should certainly be helpful to you. Thinking along often-used lines does not result in anything new. If you do not have some off-beat areas in your brain, you are unlikely to come up with innovations.

One trigger to many good minds that results in a flood of associated ideas is a "discovery." In this context, a "discovery" is a pattern or fact that is new to you. It may be old to many, many others. This is irrelevant. If it is new to you, it can be called a discovery. And, under the right circumstances, a discovery can trigger your mind so that you come up with a truly remarkable result.

Now a word of caution. All of the ideas that you get will not be good. You will discard many almost immediately. Some, after you or others have considered them carefully, turn out to have flaws. But be careful! On second thought, a "bad" idea may have some merit—or may trigger a "good" idea.

Many a good idea is simply forgotten. It comes to you in a flash. Somebody calls you on the phone, and you go off to a conference. When you get back a few hours later, it is gone. You may even scribble a few notes in haste, and then find later that you cannot read them. It happens to everyone who has ideas. The only answer is to have so many that a loss of a few is not a complete disaster.

And, finally, you may carefully write down your idea in a notebook. Then you forget it. And that is the end of another good idea. The notebooks of Newton and Leonardo da Vinci are full of good ideas that were never carried out to a useful conclusion.

The importance of timing

Often, the timing of your new idea is extremely important. It must be just right. Too soon, and your idea will not be accepted. Too late, and it may have little value.

The "climate" must be right for accepting a good new idea. As an example of the importance of timeliness, take Newton's invention of the calculus. The scientists of his day were interested in continuous variables. They were concerned about the motions of bodies, including those of our solar system. The invention of the calculus was enormously helpful in gaining understanding and making further advances. Much later in the history of physics, Planck had the ideas that led to the formulation of the laws of quantum mechanics. His theories dealt with "packets"—not with continuously-varying quantities. And his thinking about packets revolutioned twentieth century physics. Again, his ideas came at the right time. To emphasize the point, just imagine what would have happened if Planck had brought up his idea of quanta at the time of Newton. To the scientists of that day, quite likely they would have had little or no meaning. Certainly they would not have contributed to the solution of the problems to which Newton and his fellow workers were addressing themselves. For this reason, his idea might still be buried in some obscure manuscript in a medieval library. On the other hand, if Newton had not proposed his calculus when he did, then certainly Leibniz would have been credited with the invention.

Timeliness is also important to management. For when a new idea is proposed, management must always ask questions such as: Is the "climate" right? Is the market ready to accept this new product? In the process of a design that is well along, major idea changes usually are changes of philosophy. The intent of the design is changed. Often, in the course of design effort, weaknesses are uncovered. These must be corrected to insure the success of the product. But correction of weaknesses is not the same as introducing major idea changes. Any major idea changes in the course of the design must be introduced prior to the development of the subsystems. Otherwise, you just get several models of the same system. This is usually undesirable.

Simultaneous creation or invention is just another aspect of the importance of the timing of a new idea. The conception of the calculus by both Newton and Leibniz has already been mentioned. But this is far from an isolated instance. If you ever believe you have invented something and apply for a patent (or if you already have done so), most likely you will be confronted with the need to prove that you were the first one to think of your idea. Celebrated court cases have taken place between inventors, each claiming to have been the first. And many, many thousands of cases never reach the courts.

How is this situation possible?

The answer seems to be relatively clear. Two people (or two companies) both see a need. Both have all of the necessary physical and other facts necessary to come up with something new to meet the need. Quite possibly, even the patterns and arrangements are known to both— or can be found. So both come up with an answer almost simultaneously. You are quite familiar with this phenomenon in consumer goods. No sooner does one manufacturer come out with a tooth paste that he claims will result in fewer cavities than every other manufacturer of tooth pastes comes out with a competing product. And, not infrequently, the later arrivals on the scene use the very fact of their delay to claim that they have an even better and more efficient product than the first on the market.

Often you may wonder whether this necessity for the newer and better product is not at the root of the need of the advertising profession to come up with so many "new ideas." Ideas for new systems are necessary. But they are not needed as often as "new ideas" in advertising. Perhaps this difference accounts for the quite obvious differences in the types of people who thrive in the two occupations.

Classes of new system ideas

Based on a definition of "system" used throughout this book, it is possible to set up some classes for new system ideas. Clearly, a new "system" is intended to perform some new or improved process or perform an existing process in a new way.

This leads us to four basic classes of ideas: those involving *functions; inputs; outputs;* or *methods of operation.* And combinations of these four are also possible.

Consider *functions* first. A new system may perform an entirely new function or functions. It may add functions to those of an existing system. Conversely, functions may be removed because they are no longer necessary. Finally, the functions may be changed in some way. In carrying out the new idea, you may use new materials or new devices or new combinations of devices. In doing so, you may achieve better performance or lower costs or both. You may change or improve the appearance in certain cases.

As in all systems, the *inputs* may be either energy, or objects, or both. In the new system, you may use a new form of energy. You may change to new input objects. You may use less or fewer inputs or kinds of inputs, or more or more kinds of inputs. The same factors of performance, costs, and the like may lead to the new choices.

Likewise, the *outputs* of a system are always energy, or objects, or both. The new system may produce new outputs, less outputs, or changed outputs. Again, the objectives are performance, costs, and the like.

Finally, the precise method of operating an existing system may be changed with the intent of obtaining lower costs, higher efficiency, or improved products.

Because these classes of new system ideas flow directly and logically from the definition of a system and hence can only involve functions, inputs, or outputs, they are complete. There are no other possibilities.

The Polaroid camera is an outstanding example of a new idea. First, by the development of a black and white negative in ten seconds, Land did away with all of the messy processing that previously was necessary. Furthermore, the picture was available in either 60 seconds or 10 seconds—an enormous saving in time. As a result, the Polaroid camera is a "must" in every scientific laboratory. Also the speed of the film made it possible to use extremely simple and inexpensive lenses. Even more, these lenses had enormous depths of field. As a result, focusing became unnecessary: a function previously required could be omitted. All this was accomplished—and then many of these advantages were extended to color!

As another example, take the paperback book. This is a new way of performing an old function. For millions of us, paperbacks have replaced hard-cover books. They are cheaper, and they are more widely available. Although they were long a fixture in the European bookstore, their use in America is relatively recent. Obviously the paperback filled a need.

Your father probably used a straight-edged razor. These days, a straight-edged razor is hard to find. Various forms of safety razor account for a large share of the market; electric shavers, substantially all of the rest.

As one final example of a new function, take color TV. Your color set both offers you more functions (the color information) and new functions (again, the color information). These increased and new functions cost you money—but the American market is willing to pay the cost. Another "new idea" has been carried through all of the steps of system design and put to use.

As an example of a new input, take the introduction of aluminum as a cheap raw material. Today you seem to find this light metal almost everywhere. It is used for kitchen pots and pans and in airplanes. It serves as conductors for high-tension electrical lines. It is being introduced as a replacement for copper conductors in some telephone cables.

A legion of examples of new outputs is possible. Just take all the varities of breakfast cereals you find in your supermarket. The basic ingredients are still common grains: wheat, oats, corn, and rice. They still may be like flakes or puffs but they are a *different* output—at least so you are told by your TV. At one time, safety pins and zippers were truly new outputs—and welcome ones.

And, finally, examples of new methods of operation are easy to find: the cotton gin, the steam engine, the stereophonic phonograph, the electronic computer—to name just a few. All of these offer ways of doing something that people want—of meeting a human need.

Approaches to new system ideas

As a system designer, you may base your search for a new idea on a new premise or a new set of numbers. Mathematicians often use this approach. New kinds of mathematics have resulted by a change of axioms or assumptions. Non-Euclidean geometry is just one example. The "new" mathematics your children are now being taught is somewhat akin. In this case, the numbers are the same; only the point of view (the axioms) is different.

Even when you use the same old premises (or axioms), new patterns can lead to new results. Mathematicians discussing this type of approach sometimes use the words "extrapolating" and "interpolating." To them, extrapolation is the step beyond. Interpolating is filling an existing gap—perhaps one that had gone unnoticed.

System designers also use extrapolation. They build a turbine larger than any ever built before. To design it they use all of the accumulated knowledge gathered from the earlier designs. But they take a next step: they extrapolate into the unknown.

System designers also use interpolation. Your company may be producing electric motors in a standard series of sizes. Now a need arises for a motor in between two of the sizes being made. The "new" motor is not the same as either of the two motors of lesser and greater output: it is an interpolation. Perhaps the step of interpolation necessary to design the new motor is not great. Yet the result is "new." It did not exist before.

Simplifications may result from a new pattern of system thinking. The mathematicians are particularly appreciative of a new proof that is simpler and more "elegant" than previous proofs of a theorem. Sometimes a great deal of effort is devoted to obtain more elegant proofs of known theorems. Textbooks are full of the examples of the result of such efforts.

As a system designer, you may have the idea for a new process that involves fewer steps to produce a wanted output. In many ways, your thinking is somewhat like that of the mathematician seeking an elegant proof. New chemical processing systems and new manufacturing processing systems result from such search for simplification.

Finally, as a systems engineer, you may need a new pattern that involves "amplification." You may wish to produce a new or better product and can only do so if you add new processing steps or closer control for existing steps.

Amplification should not be confused with extrapolation. In this connection, as defined above, extrapolation is a taking of a new step forward. Amplification is the improvement or a new step to give better results. Of course, in a sense, these better results may also be "new."

Recognizing a need

You may search for an unfulfilled need. In fact, many companies have market research organizations to study opportunitities for new products. Also they try to keep up with what their competitors are doing.

All ideas do not come from the market research people. An engineer may conceive of a new system or an improvement of an existing system in the course of his work. In the Bell System, many of the products with which you are familiar have been the results of suggestions of their research and development people. A salesman may learn of an unmet need in talking with his customers.

In the last few decades, our government has been a frequent and persistent source of new needs. The Department of Defense approaches companies for new weapons and new weapons systems for both offense and defense. The Department of Agriculture needs new chemicals to overcome insect pests. Then need of our aerospace systems to put a man on the moon are publicized in newspapers and magazines. We are all aware of them. Even more aware are the companies to whom requests have been made for the required system developments.

Some useful techniques

What techniques are useful in getting an idea?

You want to pile up tentative ideas. You want many alternatives. At this stage, you are interested in quantity—not necessarily quality. Weeding out the poorer ideas comes later. Also, remember that facts are not in themselves ideas. However, ideas evolve from facts.

Books and learned papers about getting ideas are easy to locate. Unfortunately, almost without exception, the authors confuse getting an idea with solving a problem—and even with decision making. These are quite separate steps in system design, as you now know. To repeat, getting an idea is not a problem-solving procedure. Only in the special case of trying to get an idea for a method of attack on a problem do the two concepts overlap. But in your thinking, it is easier to treat the process of getting an idea as quite distinct from the process of problem solving. By doing this, you can avoid unnecessary confusion and can concentrate on the two techniques separately.

Although facts are not ideas, good ideas are based on facts. And some large system-oriented organizations go to great trouble and expense in their preparations for the idea stage. In fact, some organizations have Needs Research groups and Environmental Research groups who make extensive studies and analyses, and possibly even preliminary tests as groundwork in stating the problem about which ideas are needed.

In gathering facts about the needs in environment, they may distribute questionnaires; they may conduct experiments in the laboratory; and they may set up field tests in selected marketing areas. The facts gathered by the preparatory work may be useful in directing the nature of the ideas. Also they may be useful in a preliminary screening to select those that seem to be most promising in view of the indicated needs and environment.

Many excellent ideas are the sole creation of one person. Of course, this is quite possible because the human brain is the source of ideas. And for some kinds of ideas, one brain at work is worth more than a dozen.

Human imagination plays the leading part. Why? Because with your imagination you can alter axioms, change premises, try pattern after pattern, and jump to conclusions. Or get an inspiration! Then, you know intuitively that your new idea is good. It is the answer that you need. The information-processing of the human brain that leads to this kind of inspiration is probably beyond human understanding.

Somewhat akin to inspiration is serendipity: uncovering an idea when you are looking for something else. But serendipity only works if you have adequate facts and have stored in your subconscious an understanding of the need.

When you are working alone, what tools can you use? Possibly, one of the most powerful tools is *analogy*. You know about or can imagine a parallel situation and the answers that applied. Some of the facts may have been similar—or some of the pattern. Or, conversely, some of the facts were exactly the opposite. You ask yourself stimulating questions along these lines. Then you let your brain search for the answers.

One thing that you do *not* do in trying to get a new idea is use the "scientific method." You do *not* set up a set of hypotheses, and then try to work out a solution and make a crucial test. Oh no! Instead, you want a new idea. You are *not* trying to solve a problem. You are *not* trying to set up a test. Unfortunately, in school you are taught the application of the scientific method and almost nothing about the techniques of getting a good idea. Is it any wonder that most good ideas are contributed by a relatively small number of our talented people?

Ideas can also be generated by group activity. A husband-and-wife team write a book. Two researchers get together and toss thoughts backward and forward. In large scientific laboratories, small teams of two or three people are asked to explore a field. As an extreme case, an ad hoc committee may be set up. Such committees often have as many as ten or twelve people. Sometimes, a committee comes up with worthwhile ideas; often, with a voluminous and useless report.

At least in theory, you might think that two heads are better than one. Often they possess more knowledge and also more imagination. But observation and experience confirms that too much knowledge may actually be detrimental. Why? Because with more knowledge and experience often go more prejudice: knowledge of more things that won't work. More knowledge often leads to caution.

In piling up tentative ideas in the initial stage, an intelligent visitor with relatively little familiarity with the subject matter may contribute a great deal. He may come up with an unexpected analogy that opens up new vistas. On the other hand, he may not be able to contribute nearly so much in screening the ideas and solving the system problem.

A "brainstorming" group may be thought of as a special case of group activity directed toward getting many alternative ideas. The invention of brainstorming is ordinarily credited to the late Alex Osborne. He used it extensively in his creation of advertising campaigns.

When used to generate ideas, brainstorming can be quite effective. Unfortunately, other people have tried to use it as a problem-solving technique. Then, as might be expected, the results sometimes are not very good. Considerable difference of opinion on the value of brainstorming is found in the literature. Its supporters sometimes use the word "creativity" to describe its use; its detractors call it "cerebral popcorn." Both positions are tenable. Why? Because the use of brainstorming in the wrong situation is like trying to subsitute a monkey wrench for a screw driver. It just doesn't work very well.

For a brainstorming group, Osborne used five to ten people. However, he mentions larger groups—even up to 150. He believed the ideal number to be between ten and twelve. Success was had with groups consisting of all men, all women, and mixed.

Brainstorming works best when a question can be asked that has many possible answers. It should be simple to state clearly; easy to grasp; and easy to talk about. These requirements are necessary because an ideal brainstorming group has at least a few members who are not completely familiar with the problem, but can grasp ideas and carry on. To be successful in generating ideas, brainstorming must be conducted with some simple rules. Although handbooks have been written on this point, the basic principles seem to be these:

1. The problem is simple and well understood by those participating.
2. Quantity is wanted. With more ideas, the chances of good ideas increase. For this reason, any idea (however wild) is welcomed.
3. Combination and improvement are desired. Members should build on each other's ideas—including combining them to get an even better idea.
4. Suggested ideas are not evaluated.

From a psychological standpoint, brainstorming seems to have much to recommend it. As already pointed out, brainstorming groups should have more total knowledge and more total imagination than any one of its members. Competition within the group tends to stimulate the generation of ideas. Also, each member can build on what others have suggested. This, in itself, is a plus. And, finally, praise and encouragement of the members of the group act as a further stimulus on each participant.

Some of the advantages of the brainstorming group can also be used by an individual or small team. After all, you as an individual can understand the problem. You can generate ideas. You can build on your own idea. And if you are wise, you can postpone evaluation until a later time. You may lack the praise and encouragement of a group when you are working by yourself. But you also can concentrate better. And competition between individuals is not always the best way to go about generating new ideas.

You can see that there are two sides to the question: What is the best technique for generating a lot of new ideas? There is no clear-cut answer. Depending on the circumstances, one man may be enough. More would be wasteful and even hindering. Under other circumstances, a brainstorming group can be quite effective.

Getting new system ideas: a checklist

In the early stages of thinking about new ideas for a system, a checklist is often recommended as a help. But even when properly used, a checklist can only give you an initial idea or suggest a way to start. Hopefully, by some association of ideas, it may trigger your subconscious mind. As a practical matter, it often does. You get the first flash of an idea—then a flood of ideas follows.

Checklists intended to stimulate ideas are not new. In his several books, the late Alex Osborne explains checklists which have been used with success in the advertising field. Other checklists have been proposed by mathematicians—and even by some engineers. None of these seem to be particularly applicable in the area of system design. For this reason, a new checklist has been drawn up.

Table 7-1 gives the questions in the checklist.

But remember this: a checklist is intended only as a beginning. Each question in the list that you find is pertinent to your system should be expanded. Each question should stimulate your thinking. Each should be replaced by five or ten (even many more) questions. Use it as trigger for your subconscious, and quite often you will get new ideas.

Management's role: evaluation and screening

While acting as a manager, you will always be faced with a scarcity of resources. You will have alternatives for using your time and any people, money, and facilities at your disposal.

Because resources are always scarce, new ideas must undergo a preliminary evaluation or screening. By this screening the bad ideas are put aside; the good ones are selected for more careful examination.

The separation of the function of screening from that of getting a new idea is important. And the two functions should be separated in time. Get the ideas first! Screen them later. Otherwise, screening can interfere with the very objective you are seeking: getting really good ideas. For some reason, we human beings often emphasize what is wrong with an idea—not what is right.

Table 7-1.

Broad Questions

    Is the problem stated as simply as possible?

    How else could it be stated?

    Is any requirement out-dated?

    Does anything just reflect custom, tradition or opinion?

    Could this item be changed? Reduced? Eliminated?

    What could be substituted for this item?

    In what new or other ways could this be used? Of a multiuse system: What else *can* it do?

    What else *should* it do?

    Can wasted output be used?

    Use the package for something afterward?

Possible Information Sources

    Search the patent literature? Trade journals? Handbooks? Reference books? Competitor's offerings? Other systems?

    Put myself in someone else's shoes?

Likenesses and Differences

    What is similar? Can you describe it by what it isn't?

    What function can be taken from something else?

Changes and Substitutions

    Improve performance?

    Why do it this way? Another system instead?

    Some new way? How would a competitor do it? Another process?

    Turn this inside out? Upside down? Reverse?

    More of this? Bigger?

    Less of this? Smaller? Only part of it, or parts?

    Simpler? Go to the extreme?

    Add motion? Reduce or stop movement?

    Run this the other way? Is one now possibly the other?

    Reverse cause and effect? Another cause to produce this effect?

    Change the shape? Appearance?

In a large organization, top management set up the broad objectives and goals for every project. And, as explained earlier, the objectives should be listed as:

MUSTS—MUST NOTS.

WANTS—DON'T WANTS.

Also, the WANTS should be arranged in the order of importance.

With such a list available, screening of new ideas often is a fairly easy task. For example, if no capital outlay greater than $100,000 can be arranged, then a new idea that requires twice that amount can't be touched. No matter how good the idea, it must be put aside because it does not comply with a MUST objective. Again, suppose that any new idea must be exploited and marketed within a year. Now here is an idea—an excellent idea—but it will take five years to develop it, tool up, and sell it. It, too, must be put aside.

Even a modest amount of screening usually casts out most of the original entries. Only a few promising ideas remain. If necessary, further evaluation can result in putting aside some of those left from the first screening.

A final note: often, each of several alternative new ideas has promising features. Is a combination possible? Check and see. If so, you may have a "really good" new idea.

In a decision to go ahead, if necessary, management must give instructions and assignments to the people who will be involved so that the work will be continued. Alternatively, management may choose to make revisions in the idea and may indicate necessary changes in plan. Again, appropriate instructions are the action the manager takes to have the changes made.

In any event, management decides the fate of any new idea.

Problem Solving: Synthesis

Examples of synthesis problems

When you add 2 and 2 and get 4, exactly what have you done? First, you have solved a problem—no matter how trivial it is to a grownup. You have taken two facts (the numbers 2 and 2). You have combined them according to the rules of addition. And you have obtained an output (the answer, 4).

Problem solving always requires that you reorganize knowledge that you possess. You take facts or ideas and combine them in a pattern to get a tentative answer. In system design, this sort of operation is often called *synthesis*.

A discussion of a difficult jigsaw puzzle and how you go about putting it together can give a great deal of insight into some aspects of the system synthesis problem. As you know, a jigsaw puzzle consists of a collection of objects that can be interlocked to form a pattern in space. Usually the pattern has a rectangular or circular edge. When the puzzle is finally assembled, a picture is revealed. To simplify the discussion, the individual pieces will be assumed to have no particular pattern nor any colors: only shapes. This is not too unrealistic an assumption. The more difficult puzzles are designed so that the amount of information that can be gained from colors and parts of the picture composition is minimized.

When you empty the box containing the puzzle, you have from 500 to 1000 pieces of many shapes and sizes. These are mixed together in a random fashion.

The synthesis (solving the puzzle) simply consists in arranging the pieces correctly.

The rules of the game are simple: the pieces must be placed relative to each other so that they interlock properly. In solving the puzzle, you try and see whether two particular pieces fit together. No rule specifies any sequence that you must follow. You can take any two pieces and try to fit them together in any way possible.

A particular piece does not enter into the solution until it is selected for a trial. Moreover, each piece may be said to have "memory." It "remembers" the neighbors from which it was severed when the puzzle was made.

If you are trying to put together a puzzle you have never seen before, then you have a new problem of synthesis to solve. Furthermore, you have no reason to believe that this puzzle fits together exactly like another one that you solved last week. In other words, the usual crutch of analogy upon which synthesis often depends is very weak indeed.

Interestingly enough, you also have managerial responsibilities in your synthesis procedure. You decide when to start the puzzle, when to interrupt the solution for an interval and, finally, when to stop. If you get too discouraged, you may stop without ever completing the solution. In any case, you stop when the puzzle has been assembled, disassembled, and put back in its box for another try in the future.

Unless you have had some experience in solving such puzzles, you pick up a piece at random and a second piece at random and try to fit them together. Now suppose the pieces are roughly square. Immediately, you must try quite a number of combinations just to see if the randomly selected pieces fit together. The odds against finding a fit in a puzzle with 500 to 1000 pieces are enormous. So you pick up another piece and try to fit it with one of the first two. Again the odds are truly enormous against a fit. And so on.

In mathematical terms, you are working on a combinatorial problem. Also, the number of combinations is astronomical so that your chances of solving a difficult puzzle by random tries are substantially nil.

Yet people do solve tremendously complicated jigsaw puzzles. To be sure, it takes a good many hours. But still the time is only a trivial fraction of what would be required merely to pick up pieces at random and try and put them together.

How do they do it?

By using a general principle often employed in synthesis: by *factoring the problem*.

In a jigsaw puzzle, you start a factoring operation by trying to find all of the edge pieces. The complete puzzles usually are rectangular or circular in shape. If the one you are doing is rectangular, you look through the 500 or 1000 pieces, and select those with one straight edge. In doing so, you assume that the pieces with a straight edge are part of the outside border of the puzzle. You may find 100 or 150 pieces that meet this requirement. You have "factored" them out of the total number of pieces. Even more, you know that the straight edges must be arranged in such a way that they form the outer margin of the puzzle. In other words, you have further narrowed the choice of matching parts at random. But with only 100 or so parts to try and knowing that the edges must form straight lines, you soon begin to assemble the border of the puzzle. You may have overlooked a few pieces in your selection so that the complete border cannot be put together—but this is no real drawback. You will find them sooner or later.

When you have put the border together, you have made an enormous step forward. And you have made a dent in the pile of pieces to be assembled to complete the puzzle.

What do you do next?

By our initial assumptions, patterns and colors do not play any part in the solution. But you still have *shapes*. You go through the remaining pieces and divide them into five or six categories which have shapes somewhat resembling each other. If the puzzle is well constructed, you will find a considerable number of kinds of shapes. However, even a rough classification will be of enormous help. This is a further factoring operation.

Where is the best place to start further search? In the corners of the puzzle, if there are any.

Why?

Because in the corners, you know the shape of *two sides* of the piece that you must find. Except at the corners, you know the shape of only *one side* of the piece you must find. So you look for the corner pieces first. Once one of these is found, you know the shape of two sides of each of the two pieces on either side of it. And so on. With your pieces classified in rough categories, knowledge of two sides immediately restricts the number of pieces over which you must hunt. That is the advantage of factoring.

As time goes on, you will find places where you know the shape of *three sides* of the piece that you must find. This information further narrows your hunt for possible pieces. And, as you near the end of the puzzle, you will find that all of the sides of a missing piece are known. Then it is relatively easy to find it. Thus you get more information every time you locate and fit a piece. And each piece taken from a class of pieces reduces the number that you may have to try to find the next piece.

When explained in this way, solving a giant and difficult jigsaw puzzle sounds very easy. It isn't. But by systematic factoring, you ultimately come up with the answer. Without factoring, you probably have to give up in frustration.

If you can use both color and shape in looking for pieces, you can carry the factoring further and make solution even easier. But we deliberately choose to rule out this possibility in stating the problem.

As a second example, take the proof of a mathematical theorem or a problem in symbolic logic. To begin with, the pieces you will manipulate are symbols standing for particular items of information. Sometimes the symbols are conventional; sometimes you must invent your own.

You start with a stated problem. The rules you must follow are the permissible mathematical or logical operations. These rules are to be applied to the set of symbols to synthesize (deduce) a proof. And we shall assume that this proof is new—or, at least, unknown to you.

Sometimes the sequence in which the rules must be used is specified (for example, the rules covering operations inside and outside of parentheses). In general, you have no knowledge about how long a sequence of steps is necessary—at least when you begin.

To synthesize your proof, you arrange the sets of symbols or groups of sets of symbols in sequences of your choice, which you hope will achieve the results you want.

The final pattern you are seeking is the proof (or denial) of the theorem or logical problem.

One of the powerful tools in mathematical and logical problems is *analogy*. You try to remember some other problem that, in some way, resembles the one that confronts you. If you can think of one, you may be well on your way. But sometimes you can't.

Mathematical problems often involve a very long sequence of steps that must be taken in a very precise order. If the order is wrong, the proof is simply impossible. So, again, synthesis shows up as a combinatorial problem. And just as for the jigsaw puzzle, the number of possible combinations can be astronomical. Possibly—but not certainly—this observation might explain why many seemingly simple problems have defied the greatest mathematicians for decades—or even centuries.

Quite often, mathematical proofs are dependent on a clever factoring of the problem. Look through the pages of some of the technical journals. The proof often depends on the proof of one or more lemmas. In other words, what the mathematician said to himself consciously or subconsciously was, "If I can prove this, and this, and this, then I can prove my theorem and solve my problem." In other words, *he factored his problem.* Quite often in the literature, proofs of the lemmas are put in an appendix. This clarifies the presentation to the reader, but it is not the way in which the problem was solved.

There is one difference between the solution of a jigsaw puzzle and proof of a mathematical theorem. There is only one right answer for the jigsaw puzzle: a completed assembly. Quite often there is more than one way to prove a mathematical theorem. And mathematicians place great emphasis on an elegant solution. An elegant solution often means one with fewer steps to the Q.E.D. Unfortunately, the shorter and more elegant proof is often harder for an uninitiated outsider to follow. But then, few people (except professional mathematicians) regard mathematical journals as light, entertaining reading.

Now for a final example: a system synthesis problem. The parallelism with the two previous examples is striking.

Initially, you start with a stated set of goals and objectives for the new system.

The pieces that you manipulate are not the odd-shaped pieces of the jigsaw puzzle, nor the symbols of a mathematical or logical proof. They are the functional blocks that represent the subsystems, parts, etc. that together will perform the functions of the new system. As the new system is synthesized, care must be taken that these functional blocks can actually be realized in terms of hardware or people.

A set of rules must be observed. First, you know you cannot violate the basic natural laws (such as the conservation of energy and mass). But also have another set of constraints: the MUSTS—MUST NOTS and the WANTS and DON'T WANTS which the new system must meet.

The sequence in which the functional blocks are selected and tentatively allocated is sometimes prescribed. Thus, the design of a new airplane may depend on the choice of the engines to be used to power it. So, the engines are chosen first.

Again, the complete synthesis may be carried on simultaneously by different people—or even different groups of people. One group may be assigned one set of functions; other groups, other sets of functions. When such a division of labor takes place, management must make sure that the various efforts are coordinated. If the efforts are not coordinated, the system may not work; the airplane may not fly.

Often the synthesis is carried out by arranging the sets of functions and the blocks to represent them in a tabular or other graphical form. For example, in the design of a new business, the functions may be a tentative organization chart.

Just as in the case of the jigsaw puzzle and the mathematical problem, factoring is often a helpful tool in synthesizing a new system. A particular function (or set of functions) is broken off because you know a way to do it. As progress is made, more and more attention is brought to bear on those functions which need effort to find ways of performing them.

In the jigsaw puzzle, the solution tends to become easier as more and more pieces are found. Often, just the reverse occurs in a system synthesis. (Ways of performing the easier functions are found first.) Those that are most difficult to realize are left until last.

The final arrangement—the payoff—is a set of functional blocks that meets the set of goals and objectives.

These examples all show that a synthesis problem is basically combinatorial in nature. Because combinatorial problems can involve so many possibilities, a practical tool of enormous utility is simply "factoring" the problem. In other words, if by any hook or crook you can break it into smaller pieces, you have a much better chance of achieving a successful synthesis.

Once you have one answer, then it is another matter to polish it: to make it more elegant; to make it more beautiful, as the mathematicians say of an outstanding proof. To repeat, get *an* answer first. Then polish. Get the jigsaw puzzle together; speed is less important than getting the answer.

The nature of the problem

In a system design, your objective is always to innovate: to create something new. The facts may be old, but the patterns you use may never have been used before. A successful outcome is always a new idea.

Because your results are new, system design always involves *learning*. You are always breaking new ground as far as your experience is concerned. Possibly, someone else has preceded you. If he has, and you do not use his knowledge, then you have not done a good search job.

The ability to create a new system requires either innovation or serendipity. You are either able to use what some authors call your "creative" talents, or you literally fall into the right answer. As you read on, you will find that creativity actually may just be the result of some very hard mental activity. The nature of such activity and possible aids that you may use will be described in considerable detail.

Innovation is difficult.

Why?

Ordinarily, when you are given a problem you attack it with a set of rules that helped you with similar problems in the past. You may use rules for operating with ideas, visual forms, or mathematical symbols. All of these may be useful and all of them are used. However, when you are confronted with a problem of a new system design, you must select a set of pertinent facts. You must make sure you have all of them. Often you will have many irrelevant facts and ideas. These must be neglected. Then of all the patterns and combinations with which you are familiar, you will find that most of them are of little or no use. Even worse, they may lead to blind alleys.

Innovation may be difficult for quite another reason. In your classes in mathematics and physics (to mention only two subjects), you studied the material given in a prescribed textbook. The textbook first presents a set of facts and a set of patterns. At the end of the chapter is a set of problems. How are these problems made up? If they are well constructed, they gave you a set of facts somewhat different from those in the chapter itself. Then you should use a pattern that you are supposed to have learned—with, at most, a modest change in the pattern. In other words, if you can put the numbers (or facts) in the problem and choose the right formula, then you get a good grade. If you forget the formula, then you don't solve the problem. The situation is clear: facts plus pattern equal answer. Two plus two equals four.

To repeat, innovation in system design works quite differently. You have to assemble facts. You have to choose those that are essential and discard those that are not. The pattern you need for your particular synthesis is not contained in the previous chapter—it may not even be in any textbook whatever. You have to think up the right pattern. For many people, innovation is extremely difficult. You can see why.

In generating new ideas, an intelligent visitor may contribute in a brainstorming session. Even though his ideas are not to the point, they may trigger a good idea from someone else. Even if they are all off the mark, they can be screened out in the preliminary evaluation.

System synthesis requires knowledgeable people. Furthermore, they must be able to exercise judgment in the selection of the pertinent facts. And knowledge and judgment are acquired by experience.

Research experience is not always useful as a background in system design. Research is *inductively* oriented. It tries to derive generalizations or verify hypotheses. The research approach involves observation with or without a library search; hypothesis; experimentation (or more observation). One or more conclusions then follow.

The value of this sequence in system design is relatively small—except possibly in getting a new idea. Research may lead to new facts or new patterns upon which new systems may be based. Without application of these facts or patterns, it does not lead to a new system. The laser is a case in point. It resulted from imaginative research. It resulted in many devices with unique properties. But as a practical matter, the laser is often pointed out as a solution in search of a problem. To be sure, some applications have been found. Yet all of these applications together hardly justify the enormous amount of development effort that went into producing the many kinds of lasers available today. Good system applications for widespread use are yet to be found.

System design involves *synthesis*. It is *not* to be confused with analysis of systems. Analysis implies some existing phenomenon or system to be analyzed. It concentrates on techniques to separate the whole system into components or elements and to evaluate the performance of these elements as well as of the system as a whole. Here, again, quite different mental attitudes and techniques are necessary.

Because of the difficulty of many synthesis problems, a good understanding of the thinking process is very desirable. Understanding of the strategies that are available can help you improve your performance by enabling you to use the available information more efficiently. You can check for errors. But, more important, you increase your competence as a system designer.

The strategies for different kinds of synthesis problems differ in details. Yet there are underlying principles involved in all synthesis.

Except in a few isolatd technical areas, computers have not been very successful in solving synthesis problems. Wave filters and equalizers and some digital logic and sequential circuits are the outstanding exceptions. Useful computer techniques and programs have been worked out in these areas. But for most system design problems, computers are only a poor substitute for the human brain.

The system synthesis array

Fig. 7-2 is a modified version of the block diagram of a system shown in an earlier chapter. As before, the inputs to the processing unit may be energy, information energy, or objects, or even some combination. After the processing is performed, they emerge at the output.

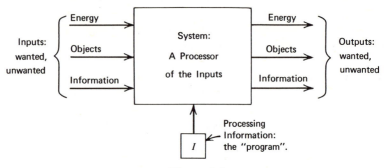

Figure 7-2.

Notice that an information input to the processing block is always required to provide instructions about the processing. Although operating information is always required, its nature depends on the processing to be performed.

Based upon Fig. 7-2, Table 7-2 can be filled out.

Table 7-2. *Kinds of System Synthesis Problems*

CASE	OUTPUT	INPUT	SYSTEM
1	G(iven)	F(ind)	F
2	F	G	F
3	F	F	G
4	G	G	F
5	G	F	G
6	F	G	G
7	G	G	G
8	F	F	F

This table contains all of the kinds of the system synthesis problems with which you may be confronted.

The operating information is supplied after synthesis of the system. For a new system, the operating information will be different from any previous case. For this reason, it has been omitted from the table, although it always is a necessary part of the synthesis task.

Note that in Cases 1, 2, and 3, one of the three components of the system are given; the other two are to be found.

Case 1 may be called the manufacturing problem. The output is wanted and some way must be found to process a suitable input to provide the output. The output may be a part, a component, or a complete system.

In Case 2 you have a utilization problem. For example, some other process may be producing a by-product. Bran in flour milling is an example. The synthesis problem consists in finding a way to turn such an input into a wanted output.

In Case 3 you have the idle machine problem. For example, you have a machine shop, and need to find work for it.

In Cases 4, 5, and 6 two of the three components are given. The third is to be found.

Finally, in Case 7 all three components are given. Expressed differently, if you start with any of the preceding six cases and reach Case 7, you have synthesized the system.

Case 8 is trivial: everything is to be found. No problem has been stated; nothing is wanted. This is the daydream situation. Or to put it another way, this is the system synthesis problem in search of an idea.

Now look at Table 7-2 from another viewpoint. The problem is solved (the synthesis is complete) only in Case 7. Suppose your particular problem falls under case 1. You search for and find a suitable input. Now you are not working on a Case 1 problem, but a Case 4 problem. You have both the input and output given. Now find a suitable functional block to perform the processing and you reach your objective: Case 7. Note that you needed two steps in your solution: from 1 to 4 to 7. Instead of this sequence, you might choose a suitable processing block. In this case you have gone from 1 to 6. Then by finding an input, you again get to Case 7.

In following either sequence 1 to 4 to 7 or 1 to 6 to 7, you may find your elbow-room limited; for example, a particular input that you would like to use may not be acceptable with certain choices for the processing block. And, of course, the reverse is also possible.

You can trace out a similar set of steps to get from Case 2 or Case 3 to Case 7. Note that one step takes you to 4, 5, or 6 and a second to Case 7.

By now, it is perfectly clear that the Given-Find array of Table 7-2 includes a representation that leads to a program of ordered steps for a synthesis design procedure. It also indicates the possible orderings for the synthesis.

After Case 7 is reached, then and only then, should you turn your attention to providing the operating information.

You know that there are several kinds of systems problems. Most textbooks do not make this quite as clear as seems desirable. The discussion of Table 7-2 holds true for any kind. But almost certainly, sooner or later you as a system designer will meet all of the possible combinations of the table.

Understanding your particular problem

Probably you have often heard the cliché "a problem well-stated is half-solved." Unfortunately, it just is not true for a difficult new system design. Also, you know it is not true for a jigsaw puzzle. On the other hand, if a problem is not well stated, you work under a handicap—often a handicap so great that a solution is impossible.

Number theory, a fascinating branch of mathematics, offers many problems that can be expressed in simple terms. Yet their solution has for centuries evaded mathematical geniuses as well as high school students. One well-known unsolved problem is Fermat's "Last Theorem." Fermat was a distinguished French mathematician of the seventeenth century. He had a well-worn copy of Dachet's book (1621) *Diaphantos*. He used it as a notebook and recorded his thoughts in the margins.

After Fermat's death, the book with his notes was published by his son Samuel in 1670: about three hundred years ago.

In the language of algebra, Fermat's Last Theorem says that the equation $x^n + y^n = z^n$ has no solution with integers x, y, z all different from zero if n is equal to or greater than 3.

If $n = 2$, then numerous examples can be found. For example, if $x = 3$; $y = 4$; $z = 5$ and $n = 2$ we have: $3^2 + 4^2 = 5^2$.

In a marginal note Fermat claimed to have proved this theorem. But despite active investigation for three centuries it has never been proved, although special cases have been. The problem is clearly stated. But it is not half-solved.

Before you start trying to synthesize a system, be sure you understand the need you are trying to fulfill. State it clearly and precisely: it defines what you are trying to do.

Set down *on paper* the essential qualities and attributes of what is wanted. Be sure you *know* what is acceptable—and what is not acceptable. These MUSTS and MUST NOTS and the WANTS and DON'T WANTS constrain your design effort.

Be careful that you check the validity of and need for each constraint. Ask questions such as: What if?

Only after you understand your problem can you start to work out suitable approaches to a solution. With understanding, you can change or modify some attributes of your synthesis to get a good design.

In your statement, cover the inputs, the outputs, and the processing functions. Leave none of their pertinent attributes out of your specification.

Table 7-3 is a useful checklist.

Table 7-3.

Broad Questions

Are *all* the *wanted* inputs shown? The outputs? The system functions?

Are *all* the *unwanted* inputs shown? The outputs? The system functions?

How about *more* inputs? Outputs? System functions?

How about *fewer* inputs? Outputs? System functions?

Make this larger? Smaller? Heavier? Lighter?

Are there any hazards? Make it safer?

Perform this system function electrically? Electronically? Hydraulically? Mechanically? Optically?

How will it work under overload? Enemy attack?

Function Modification

How can I make this easier to make? To use? To maintain? Lighter? Cheaper?

Function Location

Where else can this function be put?

What if the order were changed?

What does this go with? Next to what? On the other end? In the middle?

What about sequence? What comes first? Next?

Can these functions be combined?

What if this function were divided up? Done piecemeal?

Physical Aspects

How would I design this to be built in a home workshop?

What materials should I use? What others are possible? Is any new material available?

In what form could this be? Liquid, powder, paste, solid, vapor? Rod or tube? Triangular? Cube? Sphere?

Table 7-3. (*Continued*)

Why does it have this shape?

Would this be better symmetrical or asymmetrical?

How can a more compact design be achieved?

What if this were blown up?

What if this were lighter? Heavier?

What if this were larger? Longer? Wider? Thicker?

What if this were higher? Lower?

How can this be reinforced?

What motion is wasted?

Slide instead of rotate?

What if this were speeded up? Slowed down?

How about a longer time cycle? Shorter? Do this more often?

What other form of energy would be better?

What power is wasted unnecessarily?

How about interchangeable parts? Can costs be reduced by specifying one bolt for many places?

Are specified materials, sizes, lengths, threads, etc. standard?

Is this close tolerance really necessary?

Costs

Can I get a cost analysis of this? Where?

What is similar but costs less? Why?

Can I buy this for less? Where else can I get it as another source?

How can value be added?

Can one machine be used to make it instead of two?

Could an assembly line make it better? Cheaper?

How could it be changed for quicker assembly? Test?

How might scrap be reclaimed or used?

Appearance

What *should* it look like?

What *else* can it look like?

How could the appearance be improved?

Would another layout be better?

Should this be streamlined? Emphasized? Minimized?

What color should this be?

Can we appeal to any other sense? Touch? Soft? Hard? Smell? Perfumed?

The "black box"

With the problem clearly specified, design work can really begin. For a complex system design, a broad plan is first drawn up. Step by step, this plan is elaborated. Only when the solution is completely worked out can detailed drawings and instructions be prepared.

Almost always the first step in system design is to draw a "black box." Fig. 7-3 represents such a black box to provide any wanted operations on any inputs and outputs.

For the particular design, the next step is to list all given inputs, functional and outputs. From the system synthesis array, you can determine which the of the six possible systems problems confronts you. You know which of the possible combination of inputs, outputs, and functions are given. But you know much more, because the array deals only with the MUSTS of the problem. The contraints also furnish you with lists of the MUST NOTS. You also have a list of WANTS and DON'T WANTS.

It is important to add these at this time: you may not know what kind of input you want (for example) but *you know kinds you do not want*. The contraints limit the possible solutions and can be used to save useless effort in thinking about unacceptable solutions. For example, if you are trying to find a way to utilize a metal-working shop, you need not think of wooden objects as possible inputs and outputs. Hence, you narrowed the space over which you must search for solutions. And narrowing the space is always profitable.

After you have your black box and the lists of inputs, outputs, and functions completed, then try and find a sympathetic listener to whom you will explain what you have done.

Why?

For two reasons. (1) In trying to verbalize your problem, you may

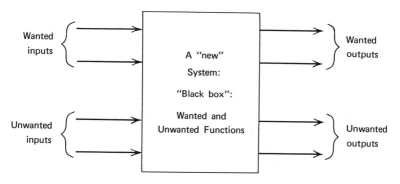

Figure 7-3.

find missing or incomplete statements. And the best time to find such omissions and commissions is early in the game. (2) Your listener may be able to make valuable suggestions. If he does, this is another advantage.

When you verbalize your problem, you tend to impress it on your own brain. For reasons that are little understood, it is possible for the unconscious to work on a problem well impressed on your brain. It is known that the unconscious can solve any problem that can be solved by the conscious mind—and many that cannot. So, in your synthesis try and get all the help that your unconscious mind can give. For some reason, verbalization can aid searches of your memory. And the synthesis of a new system is basically a memory search.

If you lack a sympathetic listener, you can still "verbalize" usefully by yourself—on paper.

As a further aid to your thinking processes, use symbols for everything possible. In geometrical problems, the symbols may be figures; in mathematical and logical problems, appropriate mathematical letters and symbols; in chemistry, chemical symbols. In system design, block diagrams. The symbols should always be simple and specific.

The use of symbols will make success more likely whether you use them to help your conscious thinking or your unconscious. In fact, it is possible—although we do not know whether it is true—that the unconscious mind may not even work with words. But the point is, you are trying to make your problem as easy to solve as possible. The use of symbols is one way to help, since they enable you to put your problem in a more compact form and thereby aid your memory operations.

For system synthesis problems, a set of symbols has already been given in an earlier chapter. These can be used as the building blocks for the system you want. You will find them most helpful.

The search for a solution

Assuming that the "black box" diagram has been completed and a set of symbols chosen, the sensible thing to do is to try to match sets of inputs and outputs using known functional operators or transforming agencies. In other words, what particular input and what particular process can give a particular wanted output?

In beginning the synthesis of a system, it pays to start by being optimistic. Take a specific example, assume that the problem falls under Case 1 of the system synthesis array: given a wanted output, find a suitable input and process. You begin by investigating whether this can be done in a single step. Usually this is very optimistic. Nevertheless, you can assume the existence of some input and then search

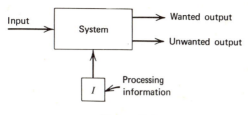

Figure 7-4.

for a process that will result in the wanted output, or you can assume some process and search for an input.

When stated in this way, it is obvious that synthesis requires logical thinking of the possibilities plus a search operation.

Fig. 7-4 represents one possible situation. Here you have input process and output (with a possibility of an unwanted output also indicated). If there is an unwanted output, you may have only a partial solution, since some further processing may be necessary to eliminate or avoid the unwanted effects. But in the simplest case, the process acts on the input to produce the wanted output.

Fig. 7-5 shows the next simplest case. Two inputs are processed in some way to provide a new combination: the wanted output. Again, the possibility of an unwanted output is also indicated.

Unfortunately, few (if any) system synthesis problems are so simple that they can be solved by the single stage represented in the figures. In real life, multiple steps are necessary. The output of one functional process is the input for the next and that output is the input for the next step, etc.

This complex situation may be represented by the combinatorial tree in Fig. 7-6. The initial input may be operated on by any one of several functional processes to produce correspondingly different outputs. In the figure, the inputs and outputs are represented by nodes and the

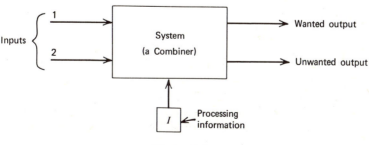

Figure 7-5.

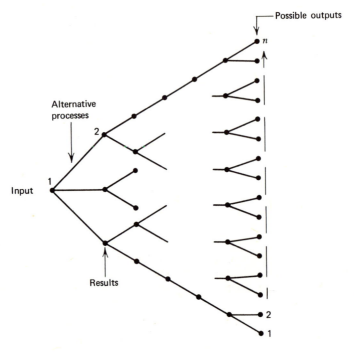

Figure 7-6.

functional processes by the lines connecting them. The wanted output is indicated in the column of nodes represented at the right of the figure.

You can see that if the first choice of process is wrong, you may never get the wanted output. Even if this hurdle is overcome and by good reasoning or some chance the initial choice is correct, then there are further chances to go astray as each additional process is chosen. And in the general case, the wanted output is only one possibility out of an enormous number. As stated before, the synthesis of a new system requires the solution of a *combinatorial* problem.

Because the number of possible combinations is so large for a complicated system, simplification is essential. And the usual name given to the simplification is "factoring."

Factoring may be helpful in two quite different ways. Suppose that in this complex system you know how to take care of some of the cases of input, processing, output. These then can be "factored" out of the entire system and set aside. The remaining part of the system can then be again drawn up as in Fig. 7-3 complete with the pertinent tables of requirements that have already been described. The number of possible combinations for the smaller problem is far less than for the entire system. For this reason, you are ahead when you use this kind of factoring. Now return to Fig. 7-5. If for any reason you are able to say that the paths that began with the input through 1-2 through 1-n are undesirable, you have immediately eliminated all of the possible output combinations containing these paths. Thus these paths have been "factored" out of the problem and need not be considered. Thus, use of the information in Fig. 7-5 giving the DON'T WANTS and MUST NOTS can sometimes be used to reduce drastically the size of the combinatorial tree. And the use of these constraints is not limited to the first processing operation. They may be used to eliminate undesirable paths at any point in the sequence.

As a skilled designer, you will want to make your job as easy as possible. Learning not to follow undesirable paths through the combinatorial tree is essential. In many ways, system synthesis is somewhat like playing a game of chess. A good chess player avoids bad plays that will not further his efforts to win the game. Each move is analyzed for three or four (or more) stages ahead to evaluate its consequences. And by experience, good chess players know that certain lines of play are not wise. In other words, they factor out these obviously bad moves in their extrapolation of the situation. As a good designer, you also factor out unprofitable sequences of operation.

One final point. Fig. 7-6 also shows why Cases 4, 5, and 6 of the system synthesis array chart may be easier to solve than Case 1, 2, and 3. For example, if you are given an input and a process that could eventually reach the wanted output, you are ahead over having to choose a process. If you had to choose a process, and chose wrong, then you could never reach the output. On the other hand, again given the input and the process, it may be impossible to reach the wanted output. This is quite clear from the figure. Stated another way, a sow's ear is the wrong input material if your wanted output is a silk purse.

Random searches

Aside from factoring, how do you go about attacking a synthesis problem?

One way is by random search: try this; try that. Change this, change that. As we observed in the example of the jigsaw puzzle, random matching of pieces without a plan can be time-consuming. And because matches come so infrequently, it can be extremely frustrating.

Yet the human brain works in mysterious ways. We get hunches. These are believed to be the result of the association of ideas in some unexplained way. Possibly, our subconscious makes try after try in much the same way as we try to assemble a jigsaw puzzle by putting pieces together at random. At any particular try, the subconscious may recognize a lack of match. Less often, it may hit upon a partial match. Finally, at the "eureka" moment, it hits on a perfect match.

Testing all the routes through a combinatorial problem takes time— lots of time. Sometimes it even takes months or years. Since the number of possible combinations in even a small problem may be millions, billions, or even larger figures, the remarkable fact is that the human brain does solve problems.

In the relatively simple game of chess, the possible number of combinations is so large that no computer has yet been able to take them all into account. Yet certain human beings do an excellent job of playing chess. And certain human beings also do an excellent job in synthesizing complex systems.

When the unconscious mind hits upon the right combination, then like a flash, the answer is there. Psychologists sometimes call this moment the gaining of "insight."

Is it possible that the insight is the result of the ability of the mind to test possible combinations against the requirements until one set matches?

Certainly, this explanation would account for the lapse of time between the statement of the problem and the moment of insight. Sometimes the time between the statement of the problem and the flash has been called the "incubation period." Perhaps a more accurate name would be the "searching period."

Recognition of a match should take only a relatively few moments. This would account for the often observed fact that insight seems to occur almost instantaneously.

Instead of random cut-and-try, the use of the reasoning powers of your brain is often more rewarding. One form of reasoning uses *analogy*. Can you think of another problem like the one you are facing? If there is a one-to-one correspondence between a problem you have solved before and the present one, then the two problems may be said to be analogous. Quite possibly, the method you used before can be applied again. Analogy can give you insight into relations and patterns. It is of little or no help in supplying unknown facts or ideas.

The success of analogy depends on having a "bag of tricks" of sound relationships at your disposal. You can only acquire such a bag through your own experience or by training.

To be useful to you in system synthesis, relationships must be so familiar that they are available for recall when needed. Frequent usage in your work makes recall easier: this is the value of experience.

Shockley has pointed out that if you have to depend on prior training you tend to remember dramatic, exaggerated, striking, unusual, or even grotesque patterns. Thus a good teacher can impress on your mind logical structures of possible value by proper presentation so that you remember them. However, a teacher can only use a shotgun technique. He cannot foresee the problems you will meet after you leave the university.

To be a competent system designer, you need both good training and the experience that comes only by trying to design. You are quite likely to remember the mistakes that you made in trying to synthesize your first complex system.

Changing the rules

In a difficult synthesis, it sometimes pays to change your attack quite radically; for example, you may go back and look at the assumptions. Can some constraint be modified? Possibly, your WANT list might be reordered and by doing so make your problem solvable.

In Fig. 7-6, you may have started in a lengthy sequence at the input of the tree. Result: a blank wall. How about starting at the output and working backward toward the input? In other words, reverse the order. Again, how about starting in the middle? Assume an input that might be derived from the input on the left; assume an output that might be converted into the wanted output on the right.

The suggestions of checking the assumptions and of changing the order of search are all quite practical in actual design work. When you are in difficulty, try them and see. Sometimes they can turn defeat into victory.

A mathematical model

If you can state your synthesis problem in mathematical form, by all means do so. In some situations, the mathematical model is *the* way to solve the problem. Unfortunately, it is not often helpful in systems synthesis. At the present time, only a few kinds of synthesis problems have been put in a form so that they can be solved mathematically.

Most mathematical texts use a set format: a set of postulates, a proof, and a theorem. This format is useful in mathematics and particularly in geometry. It seldom fits in system synthesis unless it is modified drastically. As a method, it is satisfactory where it works. But in system design, you are not trying to generate a new proof from postulates. Also the effort to remember a particular proof that you can look up is a waste of time when you are doing design work.

One advantage of a good mathematical model is the ease with which you can investigate simple cases. You can allow variables to approach zero or infinity and see what happens. And you can put the model on a computer and make all sorts of changes. Also a mathematical model enables you to incorporate the fundamental laws of invariance of mass, energy, and momentum.

In the real world, mathematical models often have a serious drawback. Very often, to make a model tractable so that answers can be computed, simplifying assumptions or approximations must be made. For example, the problem must be "linearized" even though you know that the process being modeled is actually nonlinear. If you are unlucky, the model may give completely misleading answers. Furthermore, the amount of the error introduced by the approximation may be difficult to estimate. Such errors are quite different from errors introduced by improper or incorrect mathematical manipulations. And they may be difficult to find and correct.

These comments on mathematical models may be stated in another way. Mathematics is only useful if four conditions are met:

1. The problem is so well-structured that meaningful equations can be written.
2. Adequate input information is available.
3. A method of computation is known.
4. Computations can be made in acceptable time and at an acceptable cost.

Where design methods have been known for a long time, these four conditions may be met for interpolation between existing designs or modest extrapolation beyond existing designs. For example, computers are routinely used to design certain electric power gear: transformers, boilers, motors, and generators. But note that they are only useful for modest changes in existing designs—not for the synthesis of a completely new, complex system.

A mathematical approach has been successful in the synthesis of at least two types of new systems. Network theory is mature enough so that the synthesis of wave filters and equalizers can be carried out on a routine basis by computer. A form of factoring is used so that the elements are found successively, one after another, starting from one end of the filter or equalizer. The synthesis of sequential switching circuits has also been put into mathematical form. Complex new circuits of this type can also be designed by using known algorithms and a computer.

Today, these examples are among the few exceptions that prove the rule. The design of new and complex systems by synthesis is beyond presently available mathematical techniques.

Implementation

After you have prepared a block diagram of the new system and have set down the requirements, you still must find some way to implement every block. To do this, you go through another information search. You will probably need lists of possible parts and materials, of information and power sources, and of processes.

If your preliminary block diagram cannot be realized, you have to recycle and try to find another synthesis that can be implemented.

Limitations of the human brain in synthesis

The human brain has at least two shortcomings that make the solution of complex systems problems extremely difficult:

1. The short-term, ready-access memory has only a limited capacity. At any one time, it can handle only a small number of items of information—something less than ten, it is believed. Furthermore, overworking your short-term form of memory quickly causes mental fatigue.

2. A short attention span. You can concentrate on a difficult problem for a relatively short time. The exact length varies with the individual and also depends on your physical condition and the environment in which you are trying to work.

By suitable tactics, these two limitations can be partially overcome. You can cut down on the number of items by condensing information into categories. Then, instead of remembering the separate items, you need to recall only a much smaller number of categories.

But you can go even further. By assigning a symbol to each category, you can put the information in an extremely compact form. Mathematicians are very fond of compact presentations. They can choose symbols so they can write a simple equation that represents a large matrix. And the matrix itself is a shorthand method of writing a set of simultaneous equations. In system synthesis, you can use symbols or functional blocks to help out your memory.

A short attention span is an enormous hindrance in thinking about long trains of relations. You just cannot keep all the possibilities in the mind. And yet, to solve the problem you may have to look into all of them!

What do you do?

You help yourself by using pencil and paper. In system synthesis, pencil and paper are the most important tools that you must know how to use. As you take each little step on the road to the eventual synthesis, jot it down. Then you will not forget it. You can go back and check on a point and then go forward again. Without such a record, you shortly will be lost in the maze of details that you need to remember—but have forgotten.

Some practical pointers

Learn the hard way—by experience:

1. Use pencil and paper.
2. Take some action—do anything. Start! To help you get your ideas together, a checklist may be helpful.
3. Start drawing figures, using your symbols.
4. Expect to make mistakes. You will! Revise. Improve: correct the mistakes. Work until you believe you have a solution. Then:
5. Make a second solution. Furthermore, make the second solution different from the first. By doing so, you may uncover ways to improve both solutions and come out with an even better answer.

While you are working on the synthesis, talk over your problem with other people. Often just telling them about it may give you a new idea. Discuss the facts as well as the pattern in which you are trying to work. Ask them questions. Many difficult problems are solved over the luncheon table.

A checklist

Polya, a mathematician, has summarized a set of questions to be used in solving mathematical problems. His list has been modified in Table 7-4 so that it will fit the system synthesis situation better.

Write it down!

Particularly in the intermediate stages of a design, it is desirable to store ideas, thoughts about constraints and possible troubles, and

Table 7-4.

Do you have all the facts?
Do you know about a similar problem? Its solution? Can you use the solution?
Do you know an analogous problem?
Do you know an applicable input, output, process? A combination?
Can you factor the problem? Can you solve one or more of the resulting parts?

---

preliminary results. In the early stages this storage may be quite informal—perhaps notes or brief memoranda. Later on, more formal reports, letters, and specifications are prepared. When a model is to be built, the preliminary information is put in the form of mechanical drawings and quite often instructions. And when you think about it, writing a book follows much the same pattern. You get an idea; you make an outline and some notes; you start gathering information and revising; and finally you start the actual writing.

All of these examples have been chosen to illustrate that information—even preliminary information—should be recorded (stored). It may also serve to communicate the results of the evaluation. However, many reports are made directly, man to man, without written material. In fact, in most organizations, verbal communication is by far more often used than written. For some reason, many of us do not like to write things down. For this reason, a designer may just tell a draftsman or mechanic what he wants.

Management's roles

Solving the problems of a new system is not a management function. Certainly, the top management of a large organization should not become involved in design details. However, the management must check the progress of the work and see that it meets the objectives and time schedule. Furthermore, they must check the cost of the project in terms of resources: men, money, and facilities.

In a large organization with many layers of management, the first (and even the second) line of supervision above the actual designers may be technical managers. They check on the technical aspects of the problem. The other management functions are performed at higher levels.

The low-level technical managers make sure that the patterns used in the synthesis are appropriate and that important facts are not overlooked during the design. Often they make suggestions. However, they must always be careful not to do the design work themselves. They are not designers; they are checkers of the design.

Analysis and Test

Definitions

After the synthesis of a system design comes the analysis and test of the tentative proposal. According to the dictionary, analysis is "separation of anything, whether an object of the senses or of the intellect into constituent parts or elements; as, *analysis* clarifies rather than increases knowledge; also, an examination of anything to distinguish its component parts, separately, or in relation to the whole."

As a part of the system design process, *analysis is used to predict whether the system will work* if it is built. In a sense it is a *trial* of the proposed design before a model is built.

The definition of trial is "the action or process of trying or putting to the proof; subjection . . . to a test, examination . . . to determine something in question; a trying out as an experiment to test practicability, efficacy, or the like."

In the present context, the word "experiment" has for its objective the discovery of something as yet unknown, or the verification or justification of what is already ascertained. It does not necessarily imply that anything is at stake.

At this stage in the design process, no actual model exists. Hence any experiments must be *Gedanken* or "thought" experiments. In other words, the experiments must be made by using the imagination and other faculties of the human brain.

Notice that *trial* and *test* suggest that something is being put to the proof. *Trial* is the wider *term*. Test is a decisive trial or criterion.

Whether the analysis of the proposed design is best described by the word "experiment," "trial," or "test," the purpose is to predict the performance and other attributes of the proposed system. Above all, it is intended to discover potential shortcomings or troubles.

Synthesis involves putting together things in new ways. At present, synthesis is more of an art than a science. Furthermore, it can be exciting. In synthesis, you discover; you learn. On the other hand, analysis is often just plain hard work. It may come down to a pedestrian enumeration and study of all possible cases. For example, before a space vehicle is built, an analysis is made of all of the conceivable ways in which it could fail. As soon as a weakness is uncovered, changes are made to overcome it.

In synthesis you must be an optimist. You assume that things will work out to your way of thinking. In analysis your attitude must be quite different. You're a trouble-finder: actual troubles; potential troubles; manufacturing difficulties; and operational weaknesses. You constantly ask such questions as: Is this function necessary? Are the provisions adequate? Are the margins of safety sufficient? Will the system be reliable? How about costs?

Why is analysis so important in system design?

Basically, because it is cheaper than building and testing a model. For example, by analysis, weaknesses were found in the early designs of the Boeing Super-Sonic Transport. As a consequence, redesign was started to overcome the faults. Officials of the company have said that the costs of finding these deficiencies were only about $\frac{1}{10}$ what would have been incurred if a model had been built and then changed. This is impressive, indeed. And it is by no means an isolated example.

A careful analysis is—or should be—made of every large, complex system. And for the same reason: economy of resources expended in reaching a satisfactory solution.

If you understand the concepts underlying analysis and have a systematic method for going through the process, you gain important advantages. You know where you are at any time during the process. If you are a manager, you know where the designers reporting to you are. Unless you know where you are in a process, you cannot select the relevant information and reject the irrelevant. Also, you have a framework to guide you in the handling of information to get the right answer. You can see your own errors and can develop competence.

Many years ago it was pointed out that in philosophical matters there are only four valid kinds of criticism. With some paraphrasing, you can use these four kinds in analyzing any system design.

1. One or more important facts or ideas entering into the design are wrong.
2. The facts are incomplete. Some important information has been left out of consideration.

Facts may be wrong because of incomplete information search, or by using sources containing erroneous information.

3. The logic pattern is wrong.
4. The logic is incomplete.

The synthesis may not have been carried through to a logical conclusion. Or, all possible alternatives may not have been considered.

The importance of correct patterns of relations between facts has been repeatedly emphasized. If errors of omission or commission have been made in the synthesis, then analysis should uncover them.

The value of experience

When you are looking for a "new idea," you may gain help from intelligent people who are not necessarily experts in the particular field. Not so for analysis.

An analyst is expected to find latent weaknesses in a proposed complex system. He must think of possible chances of failure that were overlooked by the people who synthesized the design. Presumably, they were competent. Hence, ferreting out potential failure modes is often difficult. Furthermore, it is not a task to be undertaken by a neophyte.

Sometimes, a great discovery is made by a young and relatively inexperienced person. Galois and Einstein are often cited as men who made outstanding intellectual contributions during their early twenties. But analysis calls for a different kind of approach and a different temperament. Trying to figure out how a new and unfamiliar complex system might fail is quite a different problem from that of synthesizing a new mathematical or physical proof—or even synthesizing a new system.

To find potential troubles, you must first understand every detail of the intended operation. But, even more, you must also have a background so that you can almost sense where troubles might occur. And such a background is best built up by being exposed to troubles in other systems.

To repeat, analysis is best done by professionals. They are best able to find any errors of omission or commission that may have been made in the synthesis. Experience can help guide them into the areas where weaknesses may lie hidden. If none are found, so much the better. But if they are found, then changes can be made to remove or mitigate them. Analysis is not a game for amateurs. The stakes are too high.

Selection of methods

Analysis can usually start by checking the function diagram or a functional block diagram. The functional block diagram is usually in the form of a single-thread representation. For such an input-output model, numbers are not usually shown.

The selection of the method of analysis depends somewhat on the objectives and goals that the design must meet. If a specification has been drawn up stating these objectives and goals, it can be used as a guide in the selection.

In the present context, *Gedanken* (that is, "thought") experiments are basically intended to test the logic of the proposed design. For a function diagram (such as that of a new business organization) they are an obvious choice. In fact, for such a diagram, a helpful mathematical model would be almost an impossibility. Thus *Gedanken* experiments are almost the only available method.

Mathematical models of a design express the relations by an equation or set of equations. If the equation(s) can be set up *and solved,* such models can be powerful tools.

To repeat, a mathematical model must meet two requirements:

1. The equations must be meaningful.
2. It must be possible to solve them. A set of equations that cannot be solved is of little value.

Some form of mathematics must be used when numbers are added to a functional block diagram. For example, computations may show how many of each type of functional block must be provided. Also, computations may be used to analyze and optimize such factors as the following:

1. Performance.
2. Feasibility (which depends on numbers and not on basic physical principles).
3. Costs.
4. Reliability, including expected meantime between failure (MTBF).

If a mathematical model is possible, it sometimes offers outstanding advantages. The symbolism is quite compact. Furthermore, the rules of relations and of manipulations can be checked for validity. Also, computations—and particularly repeated computations—with different values can give insight into important factors. This insight can lead to a better understanding of the working of the proposed system. In fact, it may point the way to a new alternative design better than that originally proposed.

Often computation can be substituted for large-scale tests that would be impossible otherwise. If successful, computation may reduce (but not eliminate) the need for tests of a physical model. Or, in some cases, it may permit making tests in the shop rather than in a field trial. A careful examination of the mathematical model may point up crucial areas of uncertainty that must be tested on a model.

To be most useful, a mathematical model should be simple. But simplicity can only be had by neglecting effects believed to be of little importance. For this reason, a model is always a compromise between simplicity and reality.

To be really useful in prediction, mathematical models should take into account all of the important variables and the relations between them. As a practical matter, this is impossible for some systems and particularly for those involving people. For many complex systems, the requirement cannot be met. Assumptions of fundamental importance may be concealed. Such a lack of knowledge may be disastrous. Even if known they may be of unknown magnitude. Again, this lack of knowledge can lead to disaster. Even more disturbing, solutions to some innocent-looking equations are beyond the reach of any known mathematical methods. Approximations to solutions are possible in some cases—but not in all. And even where approximations are possible, the amount of possible error may be unknown.

Because mathematics deals with numbers and is the "exact science," equations are often set up in an attempt to find an optimum point. In many applications, two alternate system designs may be compared in an effort to choose the better. Unfortunately, as every experienced analyst knows, the curve when computed may be so flat that it is hard to pick out a true optimum point and thus to pick the best of competing alternatives. Even worse, the use of mathematics may give downright misleading results if the assumptions are wrong or important factors are overlooked.

As an example of what you can run into, take the computation of the performance of a relatively simple system. The overall performance to be evaluated depends upon the properties of materials, devices, and subsystems that have not yet been built. The numbers to go into the equations must be "guesstimated" and guesstimates may contain gross errors. Then, so will the results of the computations.

As another example, take the prediction of costs. Here the mathematics is quite simple. But, again, much of the data may be poor. The unit cost of new parts may be unknown and again they must be guesstimated. The result? Often, the estimates of the costs are absurdly low. The costs of military hardware and spaceships show such effects quite clearly. Actual costs may be several times the estimates on which the project was authorized.

As a final example, take the prediction of the reliability of a new system. Suppose that the basic information consists of life tests extending over a few years on a modest number of samples of all of the devices to be used. Also, assume that the system is to be designed for a twenty-year life. Equations exist that permit some sort of an estimation of a reliability but they are not as accurate as one might wish. How can they be? The basic data are too scanty to be extrapolated with complete confidence. To further complicate matters, accelerated tests are sometimes used to simulate the failures of a design life of twenty years in a few weeks or months. With proper care and sufficient background, such accelerated tests may be useful. But unless carefully thought out and controlled, they may not include all causes of failure to be expected in the lengthy period of use.

To avoid the uncertainties of reliability predictions on untried devices, an extremely conservative attitude was taken in the design for deep-sea telephone cables. The objective was "no failure for twenty years." To achieve such a reliability, only devices were used for which long experience was available. And the particular devices chosen for the installation were carefully selected and given extensive preliminary tests. As a result of this cautious approach, no failures have yet occurred with the amplifiers of the system. However, fishermen and trawlers dragging the bottom of the ocean have snagged and broken the cables on a number of occasions. The breaks occurred relatively near the shore on the continental shelf, although the cable was heavily armored to prevent such occurrences. So, even with the greatest of care, some possible causes of failure may be overlooked. More recently, the inshore ends of the cables have been plowed several feet below the surface to avoid the interference of the fishing vessels with them.

After all of the *Gedanken* experiments have been made, and all of the mathematical computations have been completed, the crucial tests are made on an actual working model. The preceding paragraphs and examples were chosen to indicate why tests are so important. We human beings are not gifted with omniscience about all the properties of the materials and devices in our new system. And our mathematics are not able to take into account many important effects. Hence, we test. We test materials, devices, subassemblies, and finally the whole new system. You would not want to fly in an airplane unless you knew that every possible precaution had been taken against its failure. Flying a prototype model is one way to be sure. Even so, a few flaws may get by and have to be removed after the system has gone into service and experience has been gained.

Some useful tools

"Gedanken Experiments." These "thought" experiments are basically logical tests of the proposed design. Many of them can be put in the form: if A, B, and C, then N.

Assuming that a diagram of the functions or a functional block diagram is available, then questions such as the following may be asked:

Is the arrangement logical?
Is this block necessary?
Are a sufficient number of different kinds of blocks provided?

You check the specifications and make sure that there are no mutually incompatible requirements.

You, as an analyst, check for completeness: no necessary functions are omitted. You check that the inputs, outputs, and processes of each block are all in order. You go even further and try to foresee possible unwanted side effects.

For a given input-output set, you ask: Is it possible? Some processes are functionally or structurally not feasible.

Sometimes you find the invariance principles useful: conservation of energy, mass, and momentum.

For *Gedanken* experiments, there is no substitute for experience. If you have worked with and analyzed similar setups in the past, you can call on your stored knowledge to help you with the new design.

Finally, the end results of a careful analysis may not always be as revealing as you would like. For example, you may find that in theory a system is quite all right; in practice, it may not be. Since *Gedanken* experiments do not involve computations, you cannot say that a system is economically sound unless you go further.

Mathematical Tools. Although *Gedanken* experiments can be of great value in uncovering weaknesses in a proposed design, mathematics can give numerical results. And often numbers are important. Consider costs as a good example.

As pointed out earlier, mathematical methods are only applicable if the problem can be clearly stated in mathematical terms, and appropriate data and mathematical tools are available that permit solutions.

Analogy is often useful to suggest an appropriate mathematical model for a new system or a part of a new system. Enormous advances were made in the design of acoustical devices (such as phonograph pickups and loudspeakers) after it was shown that their mechanical performance could be represented by an equation similar to that for an electrical filter network. After the appropriate equivalent circuit was drawn by analogy and values obtained for the masses, compliances, and dissipative elements, the powerful tools of electrical network theory could be applied. Network theory rather easily lead to excellent designs that were great improvements over prior art.

Extensive tables have been compiled giving the analogs of electrical quantities for mechanical, hydraulic, thermal, and other types of systems. If you can use these analogies for your particular system design, you may find them to be invaluable tools in setting up your mathematical model. This is true because of the extensive effort that has been expended in developing electrical network theory.

Among other things you can do with a mathematical model is to examine the effects of variations and tolerances of components from their nominal values. As a particular example, take a transistor amplifier. You can assume high and low values for the power supply voltages; very good and very poor transistors in each stage; and resistance and capacitors at the extremes of their acceptable tolerances. Then you can compute the expected gain for your amplifier with any assumed combination of the variables. By doing so, you can get a lot of information and a "feel" about the expected performance of its circuit. In this way you can make a "sensitivity" analysis and determine which components must have close limits, and which can be allowed to vary more widely.

As a particular example of the use of a model, you may assume that every component is at its limit in the worst possible direction and then compute the performance of your system. This is so-called "worst-case" analysis.

In recent years, for some systems, worst-case analysis has been supplemented or superseded by a "probability" analysis. A probability analysis assumes that components have certain distribution curves around their nominal values for particular properties. By combining the distribution curves of the elements of a system properly, the effects of component variations on performance can be computed. Usually, some components will be off in the "high" direction; others in the "low" direction. So the performance actually is not as bad as if all components simultaneously had their worst values. Some transistor circuits are designed on a probability basis. The results seem to have been satisfactory. They allow the use of much wider tolerances than would have been permitted by a worst-case approach.

The analysis of a system under overload requires the examination of its nonlinear operation and, hence, involves nonlinear equations. Unfortunately nonlinear mathematics is not highly developed—at least, not highly enough so that the designer can predict the exact performance with confidence. For this reason, overload conditions are often tested on a physical model.

Many physical systems at times must operate under very heavy loads. Automobile engines, brakes, and tires, for example; bridges when several heavy trucks happen to be crossing at the same time; skyscrapers during an earthquake or a hurricane. After all, an airplane is a complex system that just barely can fly. The list is almost endless. Why? Because the constraints imposed on the design (size, weight and cost) dictate the efficient use of materials. Efficient use means under high stress: stresses so large that nonlinear operation cannot be avoided.

To determine whether some systems will be stable under all operating conditions, servomechanism theory and cybernetics may be useful. Many amplifiers incorporate feedback arrangements to improve the modulation performance and to achieve stable gain with transistor aging and power supply variations. Unless precautions are taken they can become unstable and can break into unwanted oscillation. Computer-controlled processing systems are another example. In an oil refinery, such a system may monitor and measure many quantities. Based on the observed values, appropriate adjustments are made to maintain optimum performance.

Again, for a limited category of systems, various branches of information theory such as coding theory and decision theory may be helpful. To receive information back from a deep-space experiment, very sophisticated telemetry systems are necessary. The signals back to earth from the tiny transmitter in the distant probe are very weak, indeed. To recover them from the inevitable noise that tends to overwhelm them requires careful design.

The tools just mentioned are all available for the analysis of a single-thread design. An extensive literature in the form of books and technical articles exists in each of the areas.

Further analysis is usually required to determine the performance of a complex design to take care of high-traffic conditions.

The stability of a set of components adequate to provide traffic capacity under heavy load conditions may need to be investigated. The theories developed by Lyapunov for determining the responses of such networks is a powerful mathematical tool discussed in numerous texts and articles. Unfortunately, the mathematical problems are so difficult that answers to many practical problems cannot be obtained.

To determine the performance of certain systems under high-traffic conditions, some of the work on probability and statistics is pertinent. Queueing theory is particularly useful. It enables you to investigate not only the performance under normal conditions but also under extreme overloads caused by many simultaneous inputs. Queueing theory has been applied to the waiting problems at toll booths for bridges and tunnels; to the delays of dial tone in telephone systems during the busy hour; and to the management of check-out stations in supermarkets. These are only a few examples chosen from many possibilities. For the theory has wide application.

Simulation is another powerful tool for analyzing the performance of some systems under high-traffic or overload conditions. The simulation of a system is the operation of a (mathematical) model that is representative of the system. Usually only the features believed to be important in the operation are simulated. Simulations may take the form of games—as in war games or games simulating competition between two firms endeavoring to penetrate the same market.

In simulation, the model can be manipulated in ways that would be too expensive, impractical, or impossible on actual systems. The words "computer simulation" and "gaming" are often used almost interchangeably. However, gaming is usually concerned with an experimental operational or training situation invariably concerned with studying human behavior or teaching people. The actual presence of people is not necessary to a simulation, but it is to a gaming problem. A computer is a necessity for any except the most highly simplified models of a system. Besides the physical systems with which this book is concerned, computer simulation has been used for social, political, and military systems. The literature is extensive. In all applications of simulation, the desire is to obtain numerical results. Thus they differ from the *Gedanken* experiments that test the logical structure.

The mathematical tools thus far mentioned so briefly are intended to help in the analysis of the *performance* of the system. But besides performance, predictions are required of the costs and often of the reliability. Analysis of these aspects also involves computation. However, the computations are often simpler than those required for the analysis of performance. After all, to determine the total cost you simply take the cost of item 1 times the number of such items; the cost of item 2 times the number of such items; and so on. Then add the separate costs to obtain the total cost. This is elementary mathematics, indeed.

In computing the expected system reliability, you take the expected number of failures per thousand hours (or whatever amount of time is pertinent) times the number of such items; failures of item 2 times the number of such items; and so on. Then sum these up to get the total number of expected failures. There are some niceties in converting from the total number of failures to be expected from this simple summation and the mean time between failure (MTBF). But the principles are relatively simple and can be found in many handbooks and texts.

Test of a physical model

Despite the existence of a mathematical model and of extensive *Gedanken* experiments, for an important and complex system, a physical model is always built. It may be a "skeleton" model in the sense that it may not have as many inputs, processes, and outputs as will later be used in a "full-scale" system. But every function should be present.

Why is a physical model so important? Simply because of the limitations of human beings in making *Gedanken* experiments and the weaknesses of mathematical analyses.

The proof of wanted performance is given by actually testing a model. Known inputs are applied in known numbers; the processing is observed and measured; and the resulting outputs are monitored. Unforeseen errors of omission and commission come to light in the test. Tests can give information about the performance of a model. If desirable or necessary, data on the reliability and modes of failure may also be collected. Experience gained in building a model may also clarify some cost questions and improve the estimated system costs. But this is not ordinarily the primary purpose of building a model.

Almost invariably, the results of tests of a model indicate some required revisions of the design. Of course, revisions are costly in time and money, but it is far better and cheaper to catch the weaknesses in a model than after production starts.

Performance testing of a model may be searching and extensive. It may go on for a long time. The tests of a complicated system, such as a new airplane, may require months or even a few years.

A branch of mathematics called "design of experiments" is sometimes helpful in planning the tests so that a maximum amount of information can be obtained by the fewest possible number of measurements.

A final word of caution. Even though the model performance meets all of the objectives and goals, there is still a chance of a slip-up when production starts. Because one model—or even a few models—performs satisfactorily, there is no guarantee that the production models will do likewise. To infer from past experience to future experience involves invoking inductive reasoning. And as you know, inductive reasoning is formally fallacious. To know what has gone before may give you some confidence about the future—but to be confident and to be certain are quite different concepts.

Human and computer roles in analysis and prediction

In *Gedanken* tests, as the name indicates, you use your brain to devise the tests. Furthermore, you think through the experiment. Because of the very nature of the test, computers seldom play more than a possibly minor role.

Mathematical models are devised by human beings; not by computers. A computer *may* help if you can:

1. Devise an adequate mathematical model.
2. Devise a program of instructions so that the computer can carry out the calculations you want.

A computer is essential in gaming problems and simulating the performance of a proposed design. But you must tell the computer how to react under every possible condition.

In testing a physical model, you must devise the tests to be made. In some testing situations, a computer may be programmed by you to make a series of tests in the order you want them. It may also be helpful in reducing a vast amount of data to an easily managed and more intelligible form.

To sum up, you have to do all of the planning of the analysis and testing. If you use a computer, you have to program the computer so that it can help you. If there are a large number of calculations to be made, then a computer can be a valuable tool. You can investigate more variations of a proposed design with its help. Under suitable conditions, you can find the optimum design with much more certainty. But a computer is no substitute for intelligent human planning—at least, not yet. Quite likely, it never will be.

Decision and Action

Without any question, decision making may be a difficult task. In making a decision, you choose between ways of getting something done or accomplishing some end. Quite often the decision must be a compromise between what you really want and what can be done. The best decision gets the most done and with minimum disadvantages.

A good decision is always the result of a study of alternative courses of action. And, as you have seen before, the result of an evaluation is the choice of the best alternative.

Managers make decisions in several different areas. However, most of a manager's decisions are caused by problems that arise in the course of his work. And as you have just seen, problems are an everyday occurrence in a manager's life. In fact, much of his time is spent in troubleshooting.

There are other areas in which he makes decisions that are important but in which questions come up only rarely. At infrequent intervals, he sets new goals and objectives. To do so, decision making is necessary. In setting new goals, new directions for the business may be set. Changes in the direction of the firm's activities occur only at intervals measured in years—not weeks. He may decide on new standards of performance or new price ranges for products. After several years of discussion and investigation, car manufacturers in the United States brought out a "compact" car to compete with foreign imports such as the Volkswagen. But such decisions must be made only rarely.

Kinds of decision problems

In simple logic problems, the decision is "yes" (or "no"). For example, if a is more than b then "yes." *Go* or *no go* problems are of this type. At a traffic light, if the light is red, you stop; if it is green, you go.

By some extension of this simple logic problem, uncertainty can be included. For example, if there is a 50-50 chance of a being greater than b, then the decision is "yes" (or "no").

Obviously, in these problems, a and b may be reversed without affecting the logical statement.

By a further step, the logic can take in quantitative relations. For example, if a costs more than $1 more than b then the decision is "no" (or "yes").

And it is easy to make the quantitative case take uncertainties into account by introducing probabilities of a greater than b.

You make such simple logic and quantitative decisions every day without even bothering to think about them. You buy Brand a instead of Brand b at the supermarket because, in your opinion, it is better or costs less. And you make such decisions without bothering to think of the logic and evaluation of uncertainties that result in your decision.

As a somewhat more complicated case, in a mixed decision problem, both the simple logic and quantitative cases must be taken into account. For example, you may have to choose between courses of action where the dollars to be spent or time to completion (or both) differ among possible alternatives. And, of course, the probabilities may effect each component in your problem.

And by far the most complex situation involves multistage decisions. Here a sequence of decisions must be made. For example, you make a decision. As a result of your action, something changes in the system or its environment. You observe the change. Thus you have more information available than you had when the first decision was made. So you make another decision. It causes further changes. You observe them. And so on. The decision makers who set service-station prices of gasoline make multistage decisions. They think about their price and also about the competitors' price. Each time a change is made, the environment is changed. And the competitor may change his price. His action calls for a new decision. The picture is constantly shifting.

Methods and tools

In any decision making a human operates on his input information to get a wanted result. In many cases the time available to make the decision and to take action may be short—often all too short. To a manager, it often seems that every decision is made under pressure.

All decisions involve the activity of the human mind that we call "thinking." Thus all decision making involves the information stored in your brain. You have acquired this information by your studies and your experience. Presumably you have built up a set of values and an appreciation of uncertainties. A store of such knowledge is important in your decision-making activities.

In processing the available information you may use what some people call "common sense" and others call "judgment."

Many of the trivial decisions that you make every day are automatic and without conscious thinking. You look to see if an automobile is coming before you cross the street. If it looks as though you can get across safely, you go ahead. You don't think about it consciously.

Some people make many of their decisions on the basis of a "hunch." They act intuitively and not by consciously thinking. Hunches and intuition are often right. However, at times they may be wrong. They cannot be trusted. For this reason, they should be carefully checked out before acting on them.

Some managers use other methods. For example, they may appeal to an authority; a superior in the organization, for example. Or they may call upon philosophy or ethics. For example, employment practices of many large organizations are being changed as a result of the ethical questions raised by minority groups. Again, some managers make their decisions in a purely arbitrary manner. "I want it this way, so do it." And, finally, some just fumble along using a trial-and-error approach. It is hardly logical or effective but it is the way some decisions are made.

Thus far, the list of approaches to decision making has concentrated on some of the poorer approaches. Other and more scientific approaches are available and desirable. For important decisions, logical reasoning and careful evaluation of all possibilities are mandatory.

In some cases, study of patterns of information can be very helpful. Have you ever seen a pattern like this before? Or one just the opposite? These questions can aid your reasoning.

Of course, if you can put the problem into a mathematical form, then you can draw upon the immense resources of that form of reasoning. Even more, if you have a computer available you may be able to put it to work. Many management science authorities believe that man-computer decisions are the best approach and will become increasingly important in the future. In fact, for certain types of problems, a statistical decision theory is already available. Although it only covers a small part of all decisions, it is certain to become more useful as time goes on.

A theory of values is being developed to help set up criteria for making marginal decisions. Value theory and probability theory together can be useful allies under many practical circumstances. You can see how they come into play in the simple examples given earlier of the kinds of decision problems.

Making a decision involves a careful consideration of the objectives and available resources. It also takes into account values and risks.

First, the objectives must be clearly understood. You must ask yourself:

What are we trying to do?
What is to be accomplished?
What are we trying to improve?
Where are we trying to go?
What returns do we want?

The objectives must be quite specific—not vague.
Why?
Because if they are vague you cannot use them as a standard in choosing between alternatives. They must be spelled out in detail.

As stated many times before, available resources consist of people, money, material facilities, time, and power. Questions to ask include:

What is available?
What should be conserved?
What should be used?

To the individuals who may be affected by your decision, dollars and personal satisfaction may be important. To the company, dollars, prestige, image, and competitive values can all enter into the picture. To society as a whole, legal, moral, ethical, and even esthetic values may enter into your decision. In stating your problem, you must ask yourself: Which of these values are important?

Risks and hedging

Almost all decisions involve risk. There is an adage often called Murphy's law: if anything can go wrong, it will. If you try to entertain all possibilities at once and vary all possible variables, often it could take several lifetimes to get an answer. Surely, a good manager should be able to make the right decision based on the most promising approach. But in doing so, he must discard some possibilities and take a calculated risk. In fact, some psychology research people believe that the reason some men like management is the element of risk in making decisions. But the risks should neither be too great (success should not depend largely on luck) nor too small (where anybody can win).

To try to put Murphy's law in quantitative form, you can ask yourself two questions:

How serious will it be if it happens?
How probable is it?

Now you combine the two. If it is both very serious and very likely to happen, the results can be fatal. Minor difficulties can be a nuisance. However, you may choose to ignore them as acceptable risks.

Above all, don't ignore risks. Take them into account. Let them guide you away from dangerous decisions. And if you can, minimize the effects. We shall discuss these courses of action later.

Information handling

The first step in decision making is to evaluate the form and content of the information you have. The questions you should ask yourself are:

Do I have enough good information of the kinds I need?

Is my information pertinent, complete, up to date, and accurate enough for my neeeds?

Many decision makers rely on readily available information. They may need more information, information in a different form, or different information from that usually available to them. Too much or too little information or information of the wrong kind can only lead to confusion, indecision, fear, or frustration. These are all emotional states. And a good decision should be free of emotional bias.

In making a major decision, put all the information down as it becomes available. You can't keep all the facts in your head. Write them down so that you can look at them and assess them. Always remember that your memory can play tricks on you. Don't depend on it when a lot depends on you.

Then check the information you have for completeness. And if some items are missing, you know it. You can try to get them.

With the necessary information on hand you can start to develop alternative actions. After all, the decision is intended to be a choice between ways of getting something done or some end accomplished.

In developing the possible alternatives, we trace through the same steps discussed earlier. Your objectives are set down in the form of MUSTS and WANTS. The WANTS should be ordered from most important to least important.

Then you can take each objective and ask yourself: What actions might meet this objective? In this way you can build up a collection of actions to meet the individual objectives. Then you can put them together to get a total course of action: that is, one alternative.

Don't stop with the first promising alternative. Keep on until you have several. In this way you give yourself a choice.

Next you evaluate each alternative against the objectives. You measure each against each MUST and WANT. Take the MUSTS first. Any that do not meet all of the *musts* is automatically out. Then you evaluate the remainder by comparing them with your WANTS. The alternative that gives the most of the WANTS with the fewest disadvantages is tentatively chosen. In many cases, it may offer the perfect answer. Most likely, it is a compromise: it has the fewest disadvantages.

. Now is the time to try and see whether a combination of several of the alternatives does offer a better solution than the one tentatively chosen. Often it does.

Penalties for a bad decision may be enormous. They can be thousands or even millions of dollars. They can set your whole time schedule completely awry. For this reason, your tentative decision must be checked for possible adverse consequences: some future problem that could result from taking the indicated action. You must look ahead to see if the choice is likely to have future drawbacks.

Always remember Murphy's law. Assess the possible adverse effects for seriousness. How likely are they? If the threat is too great, choose some other action.

Ask yourself:

Will this action readily and accurately do what I want?
Is the action adequately defined and clearly stated?
Will the action *really* give the desired results?
Will it produce the results *in time* to be of value?

Foreseeing a potential problem is always worthwhile. Careful examination for the possibilities of adverse consequences gives you a chance to get ahead of the game. You can sidestep possible problems or minimize their effect if they do happen.

Some managers tend to overlook the consequences of an action. Sometimes the reason lies in the subconscious. They are afraid they might turn up disagreeable or unpalatable facts. But not facing facts only puts off the evil day. Better face them before you take an action than after you have lost control of the situation.

Kinds of possible actions

Suppose you uncover a possible cause that could lead to adverse consequences. Yet you want to go ahead.

What do you do?

You try to take a *preventive action* to remove the cause or reduce its probability. You might take *contingency action*—perhaps a standby generator to offset or minimize the effects of a power failure. Or you may take a *corrective action* that gets rid of the known cause. Finally, you may be able to take some *adaptive action* that lets you live with the problem (but only if the effects are tolerable or you cannot get rid of the cause itself).

Thus, when you are faced with possible adverse consequences, you have to evaluate the possibilities and choose from these several kinds of actions.

Sometimes when you are faced with a tough decision, you can take some *interim action* to buy time until you can work out your final solution. Ingenuity in finding such interim actions is one of the marks of a good manager. By putting off a final decision—at least for a little while—you remove some of the pressure. Then, you can make the decision with more equanimity and quite likely come up with a better answer.

As pointed out earlier, the management decision may be: stop the project. Often this is unpleasant for all concerned. But sometimes there is no better alternative. The wanted performance cannot be provided; or costs are completely out of hand. It takes courage to stop a job that was started with great fanfare. But sometimes it is the wisest action. Quite a few of our defense and space development projects should have been stopped at a very early stage. You can choose your own examples. There are many candidates.

Further steps

Let us mention a few more points that you should observe before putting your plan into effect.

You must check progress against the schedule. To do this, controls and reporting procedures are necessary. You must assign responsibility for carrying out orders and make sure that the assignments are understood. Then you should follow up to see that orders are received and understood.

To detect any trouble, you should set up a check system including dates at which progress is to be reported and checked. Deadlines must be set and met. Expenditure of resources must be controlled.

A final but important chore: record the information on the problem, the alternatives considered, the evaluation of the alternatives, the decisions made, the actions taken, and the controls that were set up. Keep track of your activities, and also the results of meetings and conferences that pertained to the various steps in the decisions. Conferences are often useful because they permit a combined judgment that can lead to sound decisions.

Summary

Just knowing and understanding a design procedure is not enough: you must know useful ways to attack each step—from beginning to end. Now each step in the procedure involves thinking. But the necessary thinking activity is different for each step. Hence different approaches and techniques are necessary for each step. For example, to get a new idea you should use free-wheeling "divergent" thinking because there are no right or wrong answers. However, in synthesis, you should use "convergent" thinking because you are trying to get a solution (or a few solutions).

For each step in the procedure, several possible techniques are presented in considerable detail. Reasons for trying a particular technique are given and also the kinds of situations where it may work. Strengths and weaknesses are pointed out. Examples are used freely.

The approach is down-to-earth and practical—even though much of it is based on sound theory. For example, a table is given of all of the possible kinds of synthesis problems. Methods of dealing with them are discussed.

Several checklists offer help because they suggest possible and pertinent questions to ask yourself at various stages of a design sequence.

The relative importance of human and computer contributions in synthesis and analysis is pointed out. A computer cannot generate new ideas—only a human mind can do so.

8

Some Human Factors

Human Beings: The Innovators

The examination of the design process has brought out the unescapable fact that innovation is the result of human effort. To create a new design, a human being must reorganize existing knowledge into some new form. By introducing new ideas (or things), he breaks with the established order or custom. He creates what has never been. His activities in many ways resemble those of an artist, a composer, or a writer. For some of the steps, computers may be used as a tool. But even where computers can be so used, the human being must direct their operations; he must provide the program. Because of their importance in design processing, this chapter takes up a number of the relevant human factors in design.

In this book the word *innovation* means the complex process of *getting new ideas and introducing them into use.* Invention may or may not be involved. But getting ideas into use always requires an entrepreneur. The entrepreneur undertakes to start and conduct the enterprise, and assumes full control and risk. This is part of the managerial role. Thus two human activities, design and managing, are the vital elements in the design process.

Almost all innovation is the result of the efforts of a small number of people. All of us get answers to problems every day. And problem solving, as you have seen, is important in design. The point is: a relatively small number of people are responsible for the innovations that affect all of us. Most people just don't think of anything really new. Good new ideas and good new designs are produced by exceptional individuals. More surprising: many intelligent people just don't have the knack of getting good new ideas and carrying them through all the necessary steps.

Professor Hill of Tufts University has divided the extent of innovation into the following four levels:

1. *The Nobel prize level.* Perhaps 100 Americans can contribute pure basic research of such high quality. He notes that there were only 60 basic papers in physics from Newton's formulation of his laws in 1687 through 1933. Despite the information explosion that we read about, this number has probably not doubled since that date.

2. *Discovery and innovation: say 1000 people.* Fleming's discovery of penicillin *and* application of its properties in the treatment of disease is an example of this category. He did not throw away "spoiled" plates of his bacterial cultures. He found out why they were spoiled and drew important conclusions from what he found. He then used his conclusions.

3. *Invention: say 10,000 people can contribute on this level.* Mere issuance of a patent is not sufficient evidence that innovation actually occurred. Because innovation requires the *use* of the new idea. Many patents are never put to use. They may be impractical, uneconomical, or simply frivolous.

4. *Problem solving and decision making: say 10,000 can do this superbly.*

You hear the word "creativity" often used in connection with innovation. The common concept of creativity is likely to embrace items 2 and 3. Class 1 is only hoped for; class 4 possibly should not be included at all.

Professor Hill's classification endeavors to estimate the number of innovators in our present society. His real point is that most original thinking is done by only a few tens of thousands in our country. The rest of us—all two hundred million of us—are indebted for the improvements of our life to this handful of gifted individuals.

Seven Essentials for Success in Innovation

The path from idea to working model is a long one. To travel it successfully, you must meet at least seven requirements. Excellence in all seven is not necessarily required, but without some capability in each, your chances of success are low indeed. In other words, the following are seven *sine qua nons* for success:

1. *You must be aware of where needs exist.* Without this awareness, you cannot innovate. You must have some acquaintance with the field. You must know where there are gaps in knowledge, inefficiencies in processing, and unfulfilled wants for objects. Fasteners for clothing have been needed since Adam and Eve and the fig leaf. But innovation

in relatively modern times brought us the safety pin and zipper. What can you think of to improve on them?

In innovation, there may be no prior way of doing what you propose to do. Or you may have in mind some improvement in an existing system. However, in either case, you must be aware of an unsatisfied need or want. In the first instance, you are conceiving a new category; in the second, improvements on what is available.

Furthermore, your knowledge must be up-to-date. Take stock of yourself in these areas from time to time. New developments fill gaps; they may push back frontiers. Wants and needs change. Some needs even evaporate. Remember what happened to the buggy whip when the automobile took over.

2. *You must possess (or know how to acquire) the necessary information* about the proposed design. You need ideas, you need facts, you need relations, you need to know how things are now done. If you find you lack information, you must be prepared to (and know how to) acquire it.

3. *You must be motivated* to fill the needs. Without motivation, you accomplish little or nothing. You must possess an inner drive to want to fulfill the wants. Motivation is so important that it is discussed in considerable detail later in this chapter.

4. *You must have judgment.* You must be able to make decisions and make them correctly. One important decision is to try to fulfill the needs: to start the design process. And entering into your decision will be your own evaluation of your chances of success.

5. *You must have the necessary time* to complete and carry out the design task. And, of course, you must have whatever facilities you require to do your work. Many good ideas are never carried out simply because time is not available to do the necessary work.

6. *You must concentrate on your objectives* until the job is done. This may not be easy. Too many ideas can scatter your effort. Brainstorming at inappropriate times can cause failure. One good idea at a time has advantages. If you have more than one, donate your extras to others in your organization (if you are a member of an organization). Let them pursue them.

At times, in your design, you may encounter opposition from your colleagues or even your superiors. Sometimes their objections are valid; sometimes they are not. But where you do meet objections, you must be prepared to sell your ideas. You must communicate with the opposition.

7. When you believe you are justified, *you must persevere;* you must carry on despite difficulties, hindrances, and even opposition. Above

all, you must continually strive in spite of discouragements. Charles Kettering's philosophy is pertinent on this point. In essence, he said: "If you are a student and flunk once, you are out. But in innovation, you are almost always failing. You fail time after time—even a hundred times. You have to learn how to fail intelligently and keep on trying and failing and trying again."

In school you take a test, get a grade, and that's the end of it. Innovation is quite different. *Efficiency* in innovation is measured in terms of the time and resources you spend to get an *acceptable* answer.

To keep on trying in the face of failure requires self-discipline. The enemies of innovation are laziness, procrastination, and easy discouragement. Only if the odds are enormously against success should you consider a decision to quit.

Look back over the list and notice the importance of time factors. In large complex systems, the design process may extend over months—even years. Concentration over such large time spans may be very difficult. Yet it is necessary.

It is indeed an imposing list of requirements. Yet they must all be met if the design procedure is to be carried out.

It is difficult for one individual to meet all of the requirements. This is one of the important reasons some of our research and development laboratories assign teams to work on new designs. It is too much to expect one person to be brilliant in synthesis, objective in analysis, superb in decision making, and also imbued with enthusiasm which enables him to convince others.

Attributes of Successful Innovators

The design process should be a logical procedure. It should not be subject to human frailties of those doing the work. Nevertheless no study of the process can afford to neglect the human aspects.

An extensive literature deals with the human aspects of individuals who possess "creativity." Unfortunately, a great deal of it contains obvious contradictions; much of it is irrelevant to basic questions that should be considered. The approach taken here is to start with what an innovator is required to do and deduce the characteristics that he must possess to be able to do the job.

You may be sure of one thing: a good innovator has more than average intelligence. Yet, in the opinion of at least some psychologists, above a certain point (say an I.Q. of 120) other factors appear to be more significant than intelligence.

Because high brain capacity is essential and also a combination of other characteristics not possessed by all of us, it can be argued that great innovators are probably born—not made by training. It can be argued that the fortunate individual with a large brain capacity should be able to hold many ideas at once and call them forth so that they can be put into patterns and compared. And as you know, arranging ideas in patterns is an essential part of design synthesis.

The basic tools used in innovation are: knowledge, thinking and reasoning, evaluation and judging, and communication.

The knowledge possessed by an innovator depends on (1) the ideas he has acquired and can acquire, and (2) the patterns he knows and can find out about. This knowledge is stored both in his memory and in the books and other material that he can consult.

In thinking and reasoning, he uses reflection, imagination, and inspiration. Hopefully, he ends up with illumination: the answer.

He also thinks and reasons in making evaluations and coming to decisions.

He communicates his ideas so that they can be used; otherwise, the innovation process is not completed.

Characteristics of Innovators

Psychologically, a successful innovator is extremely complex.
Why?
Because at one stage of the design process, he must be a dreamer; then a thinker; next, a skeptic; a judge and jury evaluating alternatives; and, finally, a skilled communicator. He must be an optimist in synthesis; a pessimist in analysis. Quite an order!

Probably no one individual ever combined all of these attributes in their highest degree. But failure to possess any one of them to some degree is a serious—if not fatal—shortening.

Above all, an innovator must be ambitious; he must have a clear-cut and powerful drive to do something both worthwhile and notable.

To achieve his goals, he must be dedicated and persevering. He must have tenacity—the stubborn will to solve the problem "come hell or high water" despite any maddening difficulties and frustrations.

Of necessity, he must have knowledge and understanding of his own field. However, there is a vast difference between merely possessing a great amount of knowledge and knowing how to use it. A small amount properly used is far better than a great amount improperly used. Don't confuse erudition with innovation.

To innovate you must see relations not sensed by others. You must be able to take "leaps" that transform or connect ideas or things.

Of course, any innovator must have a good memory—at least a memory sufficient to provide him with the information he needs when he needs it. By dredging up patterns from his memory, he can arrange facts in new ways. To help in this activity, many good innovators store up old or unused ideas. Sometimes this habit is very profitable.

In acquiring knowledge, a good innovator asks questions—many, many questions. He has curiosity. He wants to know the reasons why.

He is a keen observer. He sees things that others do. But, more important, he also sees things that others do not see. And, in particular, he is shrewd in observing a need or a latent demand. Quite often, his curiosity is coupled with such a sensitivity to surroundings that he seems nervous. But this very sensitivity opens up the possibility of new experiences. And remember, you learn by your experiences—and in no other way.

Knowledge must be continually and consistently enlarged by study, by experiment, and by observation. It cannot remain static. Facts become obsolete. Some students of innovation have commented that successful innovators have a high capacity for self-instruction.

Thinking and Innovation

A characteristic of innovation is that the same mental equipment is used for many purposes. Information-processing operations are quite similar in nature no matter what the subject matter.

An innovator is usually far more interested in ideas than in people— except for the needs of the ultimate consumer of his design. Often he wants to make his ideas physical and tangible. Then he can handle and feel and look at the results of his thinking. He often has a drive— sometimes even a passion—for order and practicality.

A good innovator is conscious of dollars and performance. In fact, often—but not always—he exhibits considerable business ability. Successful artists often show the talents we have used to describe an innovator. Andrew Wyeth, Picasso, and Ted deGrazia come to mind as examples. They are sensitive to color, form, and the like. But they paint pictures that are both artistic and appealing to potential customers.

By necessity, a good innovator is an independent thinker and doer. He can't be one of the crowd. He must be a lone wolf. Often, but not always, he is a hard worker. Some innovators claim that they act the way they do because they are just plain lazy. But they are the exceptions.

Because he is an independent thinker, an innovator always has a questioning attitude and a skeptical approach to the things around him. Why do it this way? Can't we think of something better? Such thoughts continually run through his mind.

A good innovator must be adaptable. He must be flexible in his thinking. Why? Because he must be an optimist when he is synthesizing; a pessimist when analyzing. If one way won't work, he has to find another.

He must always be resourceful. Troubles always crop up. They must be overcome.

A good innovator is courageous—even bold—when that course is indicated. He must be able to act on an educated guess. He must make up his own mind despite external pressures.

To sum up, he must have a healthy ego. Innovation is not for the faint of heart.

Essentially, an innovator is a minority of one. Self-confidence is necessary to persevere to final success.

Necessary Skills

Above all, an innovator must think clearly. Intellectual competence is a *sine qua non*. But besides reasoning ability, he must possess an uncommon amount of plain common sense. In his reasoning, he must always be logical. Furthermore, in suitable circumstances, he must use mathematics expertly. And in reasoning, either by logic or mathematics, he must comply with the basic physical and chemical laws that underlie all phenomena. He must know how to use them to help himself. But he must also understand the limitations that they put on his designs.

Of equal importance with reasoning ability is imagination. With imagination, he can rearrange known ideas in a new combination. He can take facts and form new patterns. Imagination permits him to think about things that do not exist. A fertile imagination often means a copious flow of ideas. With these ideas, he can build the framework of his design. He supervises the building by using reason.

As we are using the word *imagination* in the present context, we are not talking about "fairy-tale" imagination but a practical, constructive, controlled way of using the mind. Because imagination is so important to artists and poets, a good designer and innovation may also be accomplished in these fields. Using your imagination in one field does not block its use in another. In system design, the imagination is controlled; it is orderly and practical; in the arts, it may throw off some of these restraints.

Imagination plus the ability to use it can lead to originality—even invention. With imagination, you can look beyond the obvious and arrive at original solutions. You can abandon conventional techniques and recognize basic relations long before you come up with a solution to the problem.

Besides this already formidable list of skills, a good innovator must possess a number of more specialized abilities. Among these are the ability to accomplish the following:

1. *Recognize key problems.* He must be critical of clumsy, inefficient design. He must be dissatisfied with and improve on existing things. Even his own designs must not be sacrosanct. He must woo the unconventional in his work. He must be an iconoclast, a breaker of idols. He must discard the trammels that have impeded the thinking of others.

2. *View a new problem in all its complexity.* Look beyond the obvious, if necessary, and break the problem into its essentials. Then, bring all the relevant data to bear; take facts from experience and form the required new patterns.

3. *Tolerate ambiguity in the initial stages.* Why? Because when you start a new problem, you always are faced with a puzzle. But a good innovator always feels he can solve it in some way.

4. *Conceive efficient solutions.* In doing this he should consider alternate courses. He knows (almost instinctively) that a restated problem is always a different problem.

5. *Make experiments.* Quite often, an innovator also is a tinkerer. He has considerable mechanical ability. But not always. While analyzing the design, the experiments are made in the mind: *Gedanken* experiments.

6. *Put his ideas down on paper* so that he can consider and evaluate them more easily.

7. *Analyze his design* and foresee possible troubles. Only by analysis can weaknesses be found and ways taken to forestall them.

Quite a list! But we are not through yet! For a good innovator must be willing to make decisions for himself. To learn how to do this, he must want to develop his judgment through experience. And he must have good judgment for working in the unknown. A new design always involves risks. He must understand these risks and assess their importance. For example, he must use good judgment in selecting a new design problem. Then, he must eliminate the nonessentials in planning his attack. And he must use good judgment in modifying his plans as unforeseen conditions are met.

A good innovator has complete intellectual integrity. He listens to suggestions but he makes up his own mind. He is not led astray by appearances. He is of no school; in doctrines he has no master. Often he is not a great respecter of persons but he always values things and ideas.

Truth is his primary object. Many innovators value truth so highly that they will suffer great personal discomfort and even pain to find it and testify to it.

And, finally, a good innovator must be able to communicate. He must present his ideas clearly and defend them against attack. Almost always, ideas are attacked (at least in the beginning). In his presentations, he must be persuasive. Often, he must tell what he does know, even if he does not know the whole story. In doing so, he must be careful to distinguish between what he knows and what he does not.

In communicating his ideas, he must often use graphical methods: tables, charts, memoranda, reports, even technical articles and occasionally a book. To communicate effectively in all the necessary ways, he must be assertive—willing to speak out—to tell the truth, the whole truth, and nothing but the truth.

What does it all add up to? It adds up to a human being with many unusual talents and abilities: a healthy ego; abounding self-confidence; ability to meet troubles and overcome them. Perhaps the need for this combination of attitudes and skills accounts for the fact that so few of us can innovate successfully.

Motivation for Innovation

Motivation in general

Some psychologists think that all behavior is motivated. In other words, without motivation there would be no behavior. Thus, following this lead, it follows that innovative behavior is directed to the satisfaction of human needs.

Just as the design procedure is a complex series of steps, the motives to furnish the drives for such a major effort are many and complex.

On our government's level vis-à-vis the outside world, two driving motives are concerned with survival and prestige. The attitudes of our high-level government officials are affected in important ways by the need to preserve our country and to present a good image to the rest of the world. Look at your daily newspaper. You can sense these motives behind the headlines on every front page. In its internal affairs, government is motivated both by the welfare of the people in general and also by the welfare of the bureaucracy and the jobs it provides. These two motives (and their consequences) are obvious to any observer of the world today. They are present not only in the United States but in other countries of the world.

On the corporate level, these motives are modified somewhat. True, the corporation is motivated by a desire to survive. The reason is obvious. Some corporations pay a great deal of attention to their public "image." Some do not.

In any discussion of motives, the name of Freud must come in sooner or later. In particular, he looks upon "sublimation" as the deflection (displacement or channeling) of instinctual energy from its original (libidinal) aim to some socially desirable activity or object.

The activity or object in the present instance is to be the product of innovation. This process is sometimes called *substitute formation* or *externalization.*

According to Freud, our needs for affection, approval, prestige, independence, and power are very important. These needs are sometimes called "neurotic." They may become irrational. All of them involve tendencies to move: (1) toward, (2) away from, or (3) against people. They can lead to the defense mechanisms pointed out by Freud himself.

A person's life space represents the sum of the person himself and his environment, both physical and psychological. The resulting affects are, among others, the desire for pleasure, the desire for achievement, and the desire for enjoyment. In addition, the desire to *avoid* unpleasantness.

Everyone knows that the activity of the mind is affected by the *physical* environment. Excessive heat, excessive cold, or high humidity can lead to lowered physical and mental activity.

Emotional environment is also important. You can't do your best work under tension. For example, if you are being razzed by your associates, you are likely to become upset. The importance of such effects depends on the particular individual. Also, for a particular individual, the importance varies with time.

One contributing factor in the emotional environment is carryover psychological effects: your conditioned reflexes. If you were continually corrected when you were a child, you may react badly to correction or opposition. If you were praised when you did something well when you were young, as an adult, you may react favorably to praise. In innovation, you try to minimize the unwanted effects and maximize the wanted effects. Roadblocks and unwanted effects are discussed later in this chapter.

Personal drives may be present in the conscious mind; but much more often they lie deep in the subconscious. This is in agreement with earlier statements that the seat of all design lies in the subconscious. In this connection, Maslow, a psychologist, defines a "productive creativity" that comes from the subconscious. According to him, this kind of creativity results in "poetry, art, music, invention, science, etc."

In a particular situation, behavior may be motivated by one or more drives. The three basic drives are supposed to be food, family, and fame. However, this simplification may also be expressed in terms of five other drives:

1. *The basic psychology of the individual.* In particular, sublimation: his conversion of his (primitive) energies into wanted channels of endeavor. You must *want* to innovate! Otherwise, you won't!

2. *Satisfaction.* Successful innovators always have a deep ego involvement in achieving a successful result. They get satisfaction in creating and in serving. In fact, satisfaction has been called the "psychic" income from accomplishment. They get pleasure "from doing the job."

Economic conditions are often of less importance to innovators than other factors. Of course, they must have the primary needs of life. They must live. But when these needs are satisfied, real innovators are tremendously concerned with their work and their interest in things.

Money is one of the marks of accomplishment. Consequently, some innovators use money to measure their success. However, in many large corporations, innovators and inventors are not paid on a scale comparable to that of upper management. This is an unhappy fact. But it is a fact. Those who create the things on which large corporations depend for their growth are not adequately recognized financially.

Aside from money, innovators want and will strive for prestige. They desire to achieve. In fact, they sometimes are motivated by what seems to be a "hero's drive." Quite possibly, again in psychological terms, this need for esteem and fame may be actually a defense against rejection.

3. *Anticipation.* Unquestionably, innovators are consciously or unconsciously driven by the anticipated joys of insight—of solving a problem that no one else could solve.

4. *Freedom from frustration.* Akin to the anticipation of solving the problem is the release from the frustration of not knowing the answer.

5. *Challenge.* To an innovator, an unsolved problem is a challenge to be accepted.

Some psychologists believe that these overt drives are actually manifestations of a more basic drive: anxiety. For example, desire for money, praise, prestige, affection, eminence, and the like may stem from anxiety. Little is known about this hypothesis because experiments are difficult to devise. The experiments that have been made are not too conclusive. Nevertheless, it is an intriguing idea.

Only by expressing your own individuality can you be innovative. Standing in the way of such expression is the anxiety caused by separation from the mass or herd. To innovate, you must overcome this anxiety. You cannot be permanently stopped by it. The "average man" does not attempt to achieve the individuality he needs to be an innovator. He conforms to external demands. But the true innovator resolves this internal conflict.

Enthusiasts and job holders

Because of the effects of this internal conflict, Sarton in the Preface to Volume I of his *History of Science* points out that there are two kinds of individuals: job holders and enthusiasts. In his opinion, most of the kings and emperors of history were job holders. So were a good many of the popes. They did what their jobs required. Some of them held different jobs at different times—sometimes very different jobs. In contrast, enthusiasts are anxious to do their own self-appointed tasks. They can hardly do anything else. He also says that economic conditions deeply affect job holders but make much less impression on enthusiasts. Of course, enthusiasts have to have the basic needs of life. But when their basic needs are satisfied, real enthusiasts bother only about their work or their mission.

Job holders keep things going. But enthusiasts are the poets, artists, saints, men of science, inventors, discoverers—and innovators. They are the real creators and troublemakers.

Sarton also points out that there are bad enthusiasts. These enthusiasts follow a mirage. They delude themselves and their fellow men. Only good enthusiasts are real innovators. They have the intellectual integrity that makes them pursue the truth and communicate it to others.

A successful innovator enjoys innovation. Proverbially, he will sacrifice much for opportunities to achieve the enjoyment.

The very fact that he is an innovator sets him apart from the majority of mankind. Clearly this means he has a different temperament. Otherwise, he could not disassociate himself from the crowd.

Thus far, we have discussed the motives and drives of individuals—not of groups. When you put a group of people together to perform a design, additional motives and drives come into play. Among others, three seem most important:

1. The spur of keen competition.
2. An esprit de corps to one's own team.
3. Loyalty to an inspiring leader.

Collectively these three drives are often called "morale." Of course, morale can be high; but it also can be low if conditions are not right. With high morale, seeming miracles can sometimes be accomplished; with poor morale, nothing.

In dealing with individuals and groups doing design work, management should endeavor to appeal to the constructive drives. The best way to stimulate innovation is by creating personal motivation in the people doing the work. As pointed out earlier, many innovators are motivated by personal fulfillment or "hero's drive." Such individualistic people may be ineffectual or unhappy as members of a team. In a large organization, they may be unhappy even if they are allowed considerable freedom: there are too many "no" people to veto their proposals or to pooh-pooh their ideas.

Management has the responsibility for rewarding those who make innovations. Sometimes it does little or nothing. Patents of enormous value may go unnoticed and unrewarded. Perhaps some sharing of the royalties or the savings would be desirable. After all, top managers are rewarded by profit-sharing and tax-free stock options. Such rewards are quite uncommon for those who make the innovations that contribute to the success of the enterprise.

Age and Innovation

Some studies have shown an apparent decrease in ability to innovate with age. This belief is expressed in many places in psychological literature. Such changes may come about for one or more of the following reasons:

1. Changes in the senses that diminish the powers of observation and information collection.
2. Changes in reasoning ability.
3. Changes in the amount of experience.

These three factors can and do affect the ability to innovate. To them must be added two psychological factors:

4. Changes in drives.
5. Changes in personal environment.

According to tests, the ability to identify sensory occurrences seems to improve with age. Elders make fewer mistakes than younger people. Also, as you get older, you have available more differentiated sets of sensory properties. You can use these for cues about an object.

Psychologists think it is plausible that you learn general techniques for detecting sensory qualities. Thus there seems to be no real reason why the ability to observe and understand what one has observed should diminish with age—at least not until extreme old age. After all, you may read faster and comprehend more at 50 than at 20. In fact, the ability to observe should tend to improve with age and experience.

Psychological tests show that the capacity to form logical classes actually develops with age. Thus quite likely you can observe better as you get older and also reason more clearly.

As you get older, you have stored more facts and relations in your memory. You learn from your experiences. And experience is a good teacher. Until you lose some of your memory ability with extreme old age, you should improve as you get older.

Sometimes with greater experience, there is a hardening or fixation so that more and more is taken for granted or seen as trivial. More and more appears to be obvious, axiomatic, unchangeable, or immovable. Such a change in attitude is a definite drawback in trying to innovate.

If you are a successful innovator when you are young, as you get older your drives or achievements tend to diminish. Your life's goals may be achieved by your earlier successes. You have food, family, and a measure of fame.

Finally, as you get older and achieve some success your personal environment changes. And the change often is not for the better as far as innovation is concerned. You are bothered with innumerable details about your possessions. You have to tend to your affairs: look after your life insurance, your stocks, your bonds, your property. Your children grow up and occupy the stage as individuals. Perhaps your wife wants more of your time now that she is free when the children are in school. As a result you have less time alone to think and solve and innovate.

Some excellent research studies have apparently shown that great *inventors* tend to be young. Their performance peaks at mid-career and then drops. But for many *innovators*, their greatest achievements come in life's prime years. It seems to be a fact that fewer patentable inventions are made in the older years.

Thus, you find in the literature statements such as "in general, creators are young"; "genius and the display of new things takes youth." On

the other hand, the same sources point out "leaders are old; cold calcula-
tion and executive ability takes age."

Perhaps it is true that with age comes a tendency to merely elaborate
the big achievements of earlier years. Why try for another big breakthrough when you have already made one or more?

Nevertheless, there are many exceptions to this so-called rule that
only men in the prime of life can really innovate. Some manage it
in their youth; others, when they are well beyond the prime of life. In
fact, some studies indicate that a great innovator may do better at
70 than most of his contemporaries at 35. His work at 70 suffers only
by comparison with his own when he was young. Also, cases can be
found where the decline in ability was not really great.

Lehman has made extensive studies of gifted people in various fields.
They show an early maximum in creative production in practically all
fields. The most common bracket was between 35 and 39 years but
it varied with the field. For example, some studies show that the top
age for basic innovations among theoretical physicists and mathematicians is 30 to 35; for experimental physicists 35 to 40; for biological
and medical scientists, 40 to 45. The field of philosophy is one that
requires experience and maturity: a broad view. For this reason, it
would seem that great philosophers would produce their best works somewhat later in life.

The disheartening fact about such statistics is that the man with
greater potential usually is without reputation or experience. His
achievement may go unrecognized for years. The man with reputation
and experience is less likely to produce anything basic.

Lehman lists sixteen possible causes for the apparent early maximum
age for basic innovation. Among the more important is involvement
in affairs that successful men always find necessary. In fact, in many
large laboratories and universities, a successful innovator finds the way
to the better life lies by changing to management. This is a pity because
many a brilliant innovator is an indifferent or downright poor adminis-
trator. He also mentions the inevitable decline in health and vigor
with increased age. However, this would hardly account for the maximum observed in the thirties. Many men are quite active mentally and
physically well in their fifties and sixties; some, into their seventies.
Perhaps a more important cause is a decline in the acquisition and
application of new learning. In other words, as some people get older,
they don't keep up with the times. And an innovator must work on
the frontiers if he is to make basic contributions. If he does not keep
up with the latest developments, he automatically is less likely to be
able to contribute.

An important observation contained in Lehman's studies: most important discoveries are made shortly after beginning serious work in a field. In other words, it may be just as well to be young in a field as young in years. You don't know any better; so you create. The suggestion has merit. During World War II, the engineers and inventors at the Bell Telephone Laboratories were suddenly taken off their communications work. They had to become experts on antisubmarine warfare and radar. In these completely new fields, they made outstanding contributions.

Another possiblity derives from the fact that the decline in innovative power is less among those who are inner-motivated. Those who are accustomed to directing themselves and are interested in their work, both in breadth and depth, show smaller declines than those without these characteristics.

Haefele makes an interesting point: there is a tendency to return to one's youth as you approach old age. Events and facts long since buried in the recesses of the subconscious come to the surface again. By proper direction of this resurgence, a second period of high creativity can occur at a relatively advanced age. In fact, as a result of analyzing some of Lehman's data. Haefele believes that they show that the creative achievement of a man who lives to advanced age may actually be greater than that of many men in younger age brackets. An intriguing and provocative thought indeed!

Training and Experience

Born—not made?

Quite possible, truly great innovators are born, not made. Apparently some people want to think creatively; others do not. Education can help to develop latent powers. But many highly trained persons are sterile creatively. Others accomplish results despite almost total lack of formal education.

There is considerable indication that the quantity of knowledge that a person possess is not the most important factor in innovation. Instead the manner in which he receives, processes, and uses information is the most important factor. A little pertinent information can unlock the doors to innovation.

In this connection, notice that Noel Coward had only two lessons at the Royal Academy of Music. Frank Lesser never had a music lesson in his life. Yet both men have won fame and fortune by their accomplishments. Also, Frank Lloyd Wright did not graduate in civil engineering; yet he became a great and successful architect. And, of course, Edison is another example of a great genius in innovation who had only a minimum of formal training.

Today it is fashionable to demand that newly employed engineers, physicists, and researchers have advanced degrees. But do you remember the depression days of the 1930's? You had to have a college degree to be hired as an elevator operator at Macy's Department Store in New York City. No, formal training in a particular discipline is not the complete answer to innovative ability. Take the Radiation Laboratory at Massachusetts Institute of Technology. During World War II they recruited a staff with many kinds of backgrounds: physicists, chemists, mathematicians, even sociologists. Yet these people, with no background knowledge of weapons systems, turned their talents to the development of torpedoes and antisubmarine warfare systems. They developed new radars of kinds never dreamed of earlier. After the war, almost all of the people returned to their former work.

The value of preparation

Psychological experiments indicate that sometimes preparation helps in innovation; at other times it is a hindrance. Thus the real question is: under what conditions does preparation help?

Some writers advise that you accumulate abundant information before you attack a problem. According to them, you should know as *much* as possible about how others have tried to solve the problem. They feel that it is a serious handicap not to know what is already known about it. They sum up their position: "Six hours in a library may save six months in the laboratory!"

But there is an opposing view: you should know as *little* as possible. Then you avoid being misled by past preconceptions and failures. The less you are precommitted by training, tradition, and study the better your chances are of escaping from the grooves of accepted thought. Sometimes it is actually easier to invent something new than to go through an exhaustive search to prove that it has not been done before. Of course, positive ignorance is fatal. But history supplies many instances of the advantage of a mind not fully packed with existing knowledge and past failures.

Hyman proposes a hypothesis: (1) the content of your prior information; *and* (2) your attitude toward this content when you stored it combine to determine the outcome of your work. In this connection, a positive or negative attitude toward your own ideas and those of others can greatly affect the possibilities of your solving a problem with which you are confronted.

Since too much knowledge may commit you too soon or predispose you against new untried novel approaches, the real question is "How do you use knowledge of past experiences as an aid to progress rather than be bogged down?"

Unfortunately the psychological studies give little information on this important point.

Need for experience in the particular field

As pointed out earlier, in the design process you need four kinds of information.

1. Possible building blocks.
2. Possible relations: a "bag of tricks."
3. An understanding of the design process itself.
4. The ground rules governing the particular wanted solution.

It is hopeless to try to acquire information about the needs of a future employer before you go to work for him. For one thing, when you are in the university, you are acquiring a broad training. It is not pointed at a particular employer. But a particular employer (even if it is you) has a particular set of needs. To innovate, you must find one of these needs, and you must find a way of fulfilling it.

To fulfill a particular need, you need particular items of information. It is extremely unlikely that all of the pertinent information can be supplied you while you are in training. You have to acquire it by on-the-job experience.

Finally, the constraints under which you operate are the constraints of the employer's situation. His goals, objectives, and resources govern the choice among alternative solutions. You learn these important items of information only on the job—not in the university.

Purposes of training

Education for innovation should strive for:

1. Mastery of fundamentals.
2. Discovery of the relatedness of ideas.
3. Understanding the procedures of the design process.

Of course, it may not be possible to stimulate the imagination but, at least, your education should not throttle it. Furthermore, you should prepare for the contests and challenges of the real world.

You should be taught that knowledge consists of items of information and relations between them (organization). Also you should learn that information is a tool—not an end in itself. For example, remembering particular proofs of mathematical theorems is a waste of time except for a professional mathematician. After all, they are all in the book. You can look them up—when, as, and if you need them.

You should be taught that innovation is the reorganization of knowledge. From this and the relations you know, you can achieve a newer mental picture of the area in which you are interested. But always the mental picture involves what you know or think you know.

In the future, education should make important relationships so familiar that they are available for quick recall. Then "hunches" will occur to you by some mysterious association of ideas. Thus your education should develop and perfect a significant set of patterns. These should be so familiar to you that they are easily remembered and available for use.

Above all, you should be taught that mathematical models and mathematical reasoning can sometimes contain hard-to-find errors. Many students think that a mathematical model and solution is the answer to a problem. Sometimes it is. But often it is not.

Today, with few exceptions, education stresses the acquisition of knowledge and the techniques of analysis. But this is only a part of the necessary background for innovation. In fact, the emphasis on analysis tends to stifle innovation.

A student's performance is measured by how well he feeds back the information gained from lectures and tests. He gets a good grade when he gives the right answers to problems assembled for him.

He does not learn not to be discouraged when he makes mistakes: if he makes mistakes he is penalized. Kettering has made several remarks that are particularly pertinent. For example, "An inventor is simply a fellow who doesn't take his education too seriously." Also he pointed that if a student flunks once, he is out. On the other hand, an innovator is almost always failing. He tries and fails once, twice, many times before he finally succeeds. Thus getting good grades in a set course and doing innovation are quite different endeavors. One of the biggest jobs of education is to teach how to fail intelligently and to keep on trying and failing and trying again.

In the classroom, a semester lasts only a modest number of weeks. A considerable portion of this time must be devoted to lectures and demonstrations. There is little time to gather facts for yourself, to digest them, to consolidate them, to reorganize them, and to come up with a really new idea. It remains to be demonstrated that a classroom test involving the ability to solve a "monkey-reach-for-the-stick" type of problem is valid for evaluating the ability to innovate.

Except for the dissertation required for advanced degrees, students get little opportunity to find out how innovation really must be done. Even then, many dissertations involve only an extensive library research and a culling of quotations from many sources. There is little that is really new. Writing one—or a few—dissertations may provide some background and understanding of the procedures of innovation. Again, it may not. In any case, knowledge of the procedure is seldom emphasized as the important point. And it should be.

Many courses at the university are taught from textbooks. The author gathers his material from his own researches and published material. He prepares a manuscript, sees it through the many steps before it appears in print. By the time it is distributed and adopted as a textbook, five or more years go by. Thus a new textbook reflects the state of the art five years before—not that at the time of adoption. This is a serious drawback. This fact alone makes it almost impossible for a university graduate without experience to be aware of the information he needs to innovate immediately. Even if he does research during his graduate years, and thus is on the frontiers, he must find an employer who needs exactly that research to solve some current problem. Hence, his choice of employers is narrow indeed. Also, if he is already so specialized, and the employer decides not to pursue that line of endeavor, then he has to restock his file of information and turn to another area.

It can be argued that a five year-lag in the up-to-dateness of textbooks is no tremendous handicap. Thus the Department of Defense, in its project Hindsight, traced twenty sophisticated weapons systems, including ICBM's and night-radar systems, to find the source of the ideas. Nineteen of the twenty systems (95%) were based on science thirty to fifty years old. This raises a question: Is new science not applied until it is in the text-book, taught, and a student is well along in research or development thirty years later?

As pointed out in earlier chapters, the mortality of ideas is extremely high. For example, a survey of a group of successful companies showed that about one idea in sixty was ever successfully promoted in the market place. Only three dollars of every ten dollars spent is put on the successful product; seven dollars of the ten is lost on those that do not make the grade.

When faced with these figures, management often decides not to go ahead with an idea even though it is a good one. With odds of about sixty to one against success, management must be cautious. But another important point: the manager ordinarily is older and more experienced than the innovator. He received his training and education prior to that of the youngster. Experience has taught him to be more conservative. This is just a business fact of life.

Examples can be found of young innovators who do not risk a management's "no." They go out and establish their own firm. Hewlett and Packard did exactly that. Their success has been unbelievable. Others could also be cited. But many innovators, who would be entrepreneurs, fail.

So we are back almost to where we started. How much training and experience should you have to become a successful innovator? No one really knows. You can take any position you choose and defend it.

One Man Versus a Group

How can people work best in innovation? Singly, in teams, or in large groups?

In discussing this question, you know that no idea has ever been generated except in a single human brain. Yet many of us can work better when teamed up with the right partner or as a member of a small group. In the field of musical comedy, Rogers and Hammerstein created success after success. So did Lerner and Loewe. In science, Madam Curie and her husband worked together for many years and made discoveries of enormous importance. Brattain, Bardeen, and Shockley were awarded the Nobel prize for conceiving and demonstrating the first transistor. The large research and development laboratories of today are based upon an organized, cooperative approach.

Some ideas can be worked out by one man. However, for large new systems such as those for space exploration, a *team attack* is employed. The preferred type of team attack is not completely regimented: it allows opportunity for individual contributions—some of a high order.

One important reason for group-handling of complex design and engineering projects is simply the amount of work to be done and coordinated. One person could not handle all of the necessary details in his lifetime—or even in several normal lifetimes. The number of man-hours expended on a project such as a supersonic transport, an antiballistic missile system, or an atomic power plant is astronomical.

Another important reason is the number of kinds of specialized knowledge and techniques necessary to bring such complex projects to a successful conclusion. Today, no man is a master of all knowledge. Unfortunately, examples can be found where even highly selected teams did not know enough either: a fact that merely underscores the point!

Working with knowledgeable associates can be a real advantage. Studies of how engineers and research people get their information show that most of it comes by word of mouth. Only 25% of research people go to meetings or use publications and outside information sources. Thus the more knowledgeable your associates, the better.

Despite the advantages of having people concerned with the same problem work together, some of our large research and development organizations are fragmenting. Research is done in one laboratory; preliminary development in another; and final development and introduction to manufacture in a third. Components are built in one location; assembled in another. This fragmentation tends to make more difficult the very basic advantage of working together as a team. So the final answer on individuals working alone or as members of very large teams is not yet in.

At least theoretically, a group of competent people should have more information stored in their brains than even a gifted individual. And information that is available and retrievable is a prerequisite for innovation. And the stored information may embrace both more data and more relations.

In solving a logical problem, there seems little reason to believe that a group would do much better than a qualified individual. They might be able to try more patterns in a given time. However, the rules of logical reasoning are inflexible and can be applied by a qualified person as well as by a team.

Getting an idea is the first step in any system design. In this step, groups have been used successfully. Osborn's "brainstorming" sessions brought together five to twelve people of different backgrounds. Experience has shown that such a group can produce more ideas than an individual. However, it may be fallacious that mere quantity of ideas, alone, is good. Quality must also be taken into account.

In the "buzz" sessions described in the literature, a large group of people is divided into groups of four to six. Each group is expected to come up with new ideas. The intent is to get all possible ideas. Each small group chooses a chairman. But, since the groups are small, each member can take part. Each group gets a short period—only five to ten minutes—to think of ideas. Each group then chooses its best idea to be presented to the whole group.

In the brainstorming and buzz-session groups, some members are deliberately chosen to be people with little experience in the discussion area. Furthermore, "experts" or superiors are chosen with care—if any are included in the group. The reason: they may dampen the free flow of ideas. However, an expert may supply essential information or suggest avenues for investigation.

Another way to use a group to produce ideas is to take a number of individuals working separately. Experimentally it has been found that such a group can produce more ideas than they can when working together. Also the ideas are of higher average quality.

In setting requirements for a new design, a group may be useful. After all, many factors must be considered: for instance, the needs of the market, the resources of the company, the problems of production, and the sales angles. A group may have more of the necessary information available.

In synthesizing a new and complex system, care is necessary. True innovation is often a lonely and individualistic process. In fact, it is something like giving birth to a baby. European training emphasizes the individual nature of innovation.

Design by a committee is usually not good. The alternative is to assign a project leader to head up the design efforts. He factors the problem and assigns the various parts to groups of designers reporting to a supervisor. In this way, everyone does not work on everything. This factoring method is employed by our large research and development laboratories. Furthermore, once a group has received an assignment, the supervisor again factors it and assigns each individual his small piece of the total job.

Analysis is intended to make sure that a proposed design will work. In this area, a "tear-down" session can be useful. In essence, it is brainstorming in reverse. A group is convened to think of all the possible limitations and weaknesses of a design proposal. Presumably the group may have more pertinent experience and therefore may do a more complete job than any individual.

However, some individuals have almost a sixth sense so that their intuition is able to point out the weakness in a design proposal. Nevertheless, intuition is not always trustworthy and must be checked. In analysis, whether the work is done by an individual or a group, the attitude should be critical but fair. The intent is to spell out potential troubles so that they can be minimized or eliminated. In the same spirit, almost all technical journals have referees who read submitted papers with the idea of catching errors before the articles are published.

In decision making, a group may weigh uncertain factors better than an individual. But government by committee is not always well thought of. Probably in most organizations, one executive makes all important final decisions.

Unquestionably, large laboratories have their drawbacks. For example, your team may only have a very small problem compared to the whole project. If it succeeds, it is a team success. Then, where is the joy of accomplishment for you, a member of the team? Sometimes this is a tough question. If it is not answered satisfactorily, loss of individual motivation is certain.

To minimize this loss, some fundamental factors must be present:

1. A clear assignment of the task to be performed.
2. Confidence in the leader of the team.
3. A clear reward of some sort.

Of course the leader's freedom of action may be limited, since he too is an employee and must follow the policies set by top management.

If you are acting as an individual to carry out a design process, then you must be a designer; and you must also be a manager. And you must do both passably well.

With more than one person on a problem, one can play the role of designer; one of manager. Alternatively each can play some mixture with carefully spelled-out assignments. Obviously, more than two people can have trouble if they get in each other's way. On the other hand, if the working arrangements are smooth, they can complement each other.

In some teams, the leader or manager may feel that he knows all the answers. Sometimes he does; sometimes he doesn't. If he doesn't and then pulls rank to have his way: look out! Disaster can ensue.

One final point. Studies indicate that unless precautions are taken, even well-performing groups decline in their achievements after working together a few years. However, the decline is less if the members become cohesive and intellectually competitive. Then the experience of working together pays off. A leader must always look out for evidence of a decline in productivity of his team. If it occurs, then changes are in order.

Roadblocks to Innovation

Classification of roadblocks

In the long and complicated design procedure, many roadblocks are possible. Not only are they possible but they actually do occur. To round out the picture, the rest of this chapter discusses various kinds of roadblocks.

So many roadblocks are possible that a good grasp of the potentialities can only be had by some method of classification. Although several classifications can be found in the literature, this discussion will make the basic dichotomy of:

1. *Inside blocks.* Those within the brain of an individual which hamper his innovation.

2. *Outside blocks.* Those contributed by his training, experience, environment, and place in society.

Each of these two categories can be (and will be) divided further.

Inside blocks

Limitations of the Human Brain. No one knows whether the memory capacity of the human brain can be exhausted. But it is quite certain that the processing capacity of the brain is definitely limited. Numerous tests show that even a skilled person can handle and process only about seven independent items at one time. This is a fundamental limitation. If he is confronted with more items, he forgets, he makes mistakes. He has to go back and repeat. Now, many problems need more independent elements than can be handled within such a short-term memory span. The only way to overcome this limitation is to supplement the memory span of the brain with some external memory. In earlier chapters, this was pointed out. Furthermore, putting the problem and its factors and steps in the solution on paper was recommended. In this way, a limitation of the human brain can be partially overcome.

Since even the most talented human being has only limited intellectual capacity, he is handicapped in tackling difficult problems. If his brain capacity is below normal, obviously he is under an even greater handicap.

Aside from limited intellectual capacities, limited knowledge can be a roadblock in the way of innovation. Thus, lack of knowledge of the laws of thermodynamics resulted in the invention of many useless perpetual motion machines. If you do not have the essential facts about what is and is not possible, you are almost certain to run into a roadblock in your attempts at innovation.

It is also possible to know too much. You may be confronted with such a mass of information that it is difficult or impossible to sort out the trivial from the essential. Pages and pages of information can only confuse you when you need only one sentence hidden in a whole volume. Also, as already pointed out, if you know about many failures to solve your problem, this knowledge may be a roadblock.

Perceptual Blocks. Arnold, in his excellent study and classification of roadblocks, has pointed out the importance of perceptual blocks. Such a block results in an incorrect interpretation of the real world because of some predetermined expectation. There are several aspects of this kind of block. One aspect is that an important point is seen as obvious or trivial. The real problem is not even recognized. There may be difficulty in isolating the problem and defining it or on the contrary, the problem may be narrowed too much. There may be inability to define or isolate important attributes. In other cases, there may be a failure to use all the senses in observing (faulty observation). Yet another perceptual fault is the failure to distinguish between cause and effect.

Emotional Blocks. Emotional blocks can arise because of feelings attending the process of innovation. This is another item in Arnold's classification.

Possibly one of the most common emotional blocks is plain laziness. A person may be unable to make himself do what is necessary (or take the extra trouble needed) to do a really good job. Personal experience or observation may convince you that the sheer mechanics of writing can block innovation. Many engineers and designers find it extremely difficult to put their thoughts in writing.

In an organization, many creative people cannot escape unattractive work. Their assignments are not of their choosing. They are often given tasks that do not interest them or may even by distasteful. If they do not force themselves to become interested they will neglect them indefinitely. If interest were not aroused by anxieties (possibly fear for their job) as well as by pleasurable emotions, much work in large laboratories would never get done. But this is not efficient operation.

Other emotional blocks include fear, dread, hate, love, and the like. Some examples of these blocks are:

1. *Fear of making a mistake.* Often this is a fear of error and of failure which may destroy hard-won security.

2. *Fear of supervisors and management.* This again may come back to a fear of loss of job security.

3. In an extreme case, there may be a *pathological desire for security.* This results in a complete lack of desire to pioneer or gamble.

4. For somewhat the same reasons, there may be a *distrust of associates and subordinates.* The desire to conform is an important motive.

5. *Fear of ridicule.* If you appraise your ideas to see if others will consider them acceptable, then you are putting a block in the way of your innovative ability. From time immemorial, almost every discovery, every advancement, has been made amidst ridicule and strife. It seems that you must suffer a little if you are to progress.

6. Over motivation may result in *restricting thoughts to certain areas* that appear to be highly relevant to the problem at hand. This narrows the horizon. Overmotivation (often combined with too much pressure)

means too much speed and giving up the trial and error which gives the true feel and texture of the situation. There is no opportunity to do really basic thinking. There may be an attempt to use more information than is really available. For example, a generalization may be made from a single instance.

7. *Habit transfer and preconception.* One form of habit transfer is the carryover of past conditioning of thoughts and actions to new problems (a fixed approach). New problems are attacked with the same old methods.

All of us are conditioned by the ways we previously acted toward goals and problems. This conditioning may block approaches not tried before but containing good solutions.

Akin to habit transfer is functional fixations. Familiarity establishes a fixed usage of objects or concepts. Such fixations can interfere with innovations if the familiar object must be used in a novel way (for example, a coin as a screwdriver). It is particularly hard to change after just using an object in the usual way. In innovation, a person may misguidedly cling to a false premise or assumption concerning the task. Some causes of fixation: starting with an implicit but wrong premise; failure to see how to use an object in a novel way because it's imbedded in a conventional context; unwillingness to accept a detour that delays the achievement of the goal. Availability of a habitual response may make it very difficult to break with habit and approach a problem in a new way.

If insight is an essential element in intelligent innovation, fixation is one of its arch enemies. It is a central problem.

8. *Preconception* may take the form of wishful thinking and false associations. Just because one quality of an object is good, it does not follow that all are. Just because a girl is beautiful, it does not follow that she has a high I.Q.

9. Akin to preconception is *credulity:* the tendency to believe something is good because it is new and striking. This form of perceptual block is the inability to distinguish between the attributes of newness and usefulness.

10. Still another form of this type of block is *superficiality:* the acceptance of something because it looks good without checking to see that it really is good.

11. In some ways, *learning too much* is the opposite of credulity. The mind may be made up too soon. New and untried novel approaches are blocked off. An old picture can hamper the acquisition of a new and better picture. Yet the new cannot emerge without the existence of the old. An important innovation may be held back by a strong

commitment to an old idea. But remember that innovations always use earlier knowledge as stepping stones.

12. *Overspecialization,* learning too much about a narrow field, may limit the horizon and therefore block interdisciplinary concepts of great value.

13. *Practical mindedness.* A straight-to-the-point person gets down to facts immediately. He does not roam imaginatively around the problem. He approaches the problem too soon. He particularizes too soon. Thus he blocks out the possibility of thinking of a novel approach.

14. *Refusal to use available information.* Believe it or not, some individuals are unwilling to use what information they have because they cannot tolerate ambiguity. They can't go ahead until everything is spelled out in black and white. They need a complete specification for their problem. This too literal interpretation of the letter of the law is obviously a shortcoming—if not a complete block to any real innovation.

15. *Group thinking.* A group thinker follows the crowd. He accepts beliefs without examining them for himself. One form of group thinking is dependency on authority. Such a person is so impressed by the judgments and approaches of recognized authorities that he immediately accepts their leadership. He fails to develop any leadership of his own. Consequently he fails to innovate.

16. *Unacceptance of the new* is an often-encountered block. It partially results from the desire to conform to the group pattern. Remember, an innovator thinks alone. He is always outside the group. Otherwise, he would not be an innovator.

17. *Self-discouragement.* Since a would-be innovator meets resistance from everyone, he is often prone to self-discouragement. If carried too far, it will put an insuperable block in the way of innovation.

18. *Passivity.* This is evidenced by a lack of enthusiasm and drive to achieve a goal. Possibly it is the most detrimental block of all because, without motivation to overcome obstacles, success is almost impossible. Passivity may tie back to a desire to avoid self-evaluation or to a supposed lack of reward if success is achieved.

Many of the roadblocks in the list are due to psychological or neurotic problems. They not only may block the solution of a particular problem; they may even lead to mental breakdown such as severe depression. Or they can cause compulsive repetition of investigations which were adequate the first and second time. Persons so afflicted may be unable to complete anything. Instead of innovating, they may be unable to finish the job at all.

Outside blocks

Our Educational System. Children exhibit a questioning spirit. But in the course of going through our educational system, it becomes submerged or inhibited. A questioning spirit is seldom apparent in older people because, as children, they ran against a stone wall in grade school, in high school, and in college.

Questioning is discouraged by teachers for various reasons. They believe you should take in information and memorize it so that you can reproduce it at examination time. In solving the examination problems, you use all that is given—no more and no less—to get one right answer. You study literature very critically; you analyze it carefully. The creative side of writing is played down.

Imagination is also suppressed. Yet it is the basis of innovation. After a complete formal education you may well know too much to ever consider a certain approach to an unusual situation. Why? Because you have been taught not to question basic premises and dicta.

Furthermore, you are taught to avoid failure and worship success from your earliest school days. This early training sticks with many of us all our lives. And yet failure, failure, failure often precedes final success in innovation. You don't use only authority-given information. You don't think by rules and clichés in innovation.

Cultural Blocks. These arise from the influence of people and groups of people (society) upon the individual. Among the common cultural blocks are:

1. *Pressure to conform* to accepted patterns. Such pressure prevents a free interpretation and extrapolation of experience. Also the established cultural outlook makes it nearly impossible to see possible improvements that might be brought about by innovation. Social taboos may limit thoughts. Yet we may not even question why the limit was established. It is simply there: accept it.

2. *Overemphasis on cooperation*—on being an "organization man."

3. *Belief that indulging in fantasy is a waste of time.* As you know, imagination is the gift of the gods that leads to new things. You must dream before you achieve.

4. *Too much faith in reason or logic.* The best solutions are not always arrived at by conscious use of logic. They come, as they did to Archimedes, in a flash that leads to the cry "Eureka."

5. *Society's reluctance to accept change.* Ideas that have profoundly influenced mankind's advance often have been ignored for decades—or even generations. Finally, they are accepted. For example, Goddard

in 1907 was only an undergraduate college student. But he predicted the future of rockets which has come to pass in recent years. He predicted solid and liquid rocket fuels, a high-efficiency rocket gun, and a multistage rocket. No scientific or engineering society ever invited him to discuss his ideas before their national forums. He published a few papers—which were promptly forgotten. Half a century had to pass before his pioneering efforts were appreciated. He was ahead of his time. Society was not willing to listen to his ideas. And make no mistake, Goddard's experience was by no means an isolated one. Pulse code modulation was patented in 1920 by T. M. Raney for picture transmission. In the 1930's, H. H. Reeves also patented pulse code modulation for speech transmission. How long did it take for these basic ideas to be accepted and used? Only until the 1960's. And then only for a single system of very limited application. Today, it is being described as the coming transmission method using laser beams carried through pipes from city to city across the country.

Blocks Due to Environment. A common block to many kinds of mental work is worry caused by the pressure of daily life. Worry is the enemy of any sort of mental work—including innovation. It does not matter whether the pressure comes from off-the-job problems or on-the-job difficulties. A feeling of harassment, a mountain of details— these are some of the work pressures that cause work worry.

When it comes to innovative effort, there should never be a discouraging word. Innovation is a delicate flower that needs to be nurtured. Praise makes it bloom. Discouragement nips it in the bud.

Wisecracks can be poison. There is almost a standard list of "killer phrases" for new ideas. Any of them will work under the right circumstances; but some are more deadly than others.

Possibly the most devastating phrases are:

Has anyone ever tried it?
We tried something like that years ago.
It won't work!
That's superficial.
That's ridiculous.
That's too obvious to be considered.
That's too radical.

And then there's the old cliché. For instance:

Let's form a committee to consider it.
That's interesting but we don't have the time . . . or the manpower.

You get the idea. You could add some more such phrases yourself.

Over the long run, a lack of reward for innovative effort can become a serious roadblock. Positive rewards are promotion or recognition—or even money. Some negative rewards: produce, or be demoted, or fired.

In research and development laboratories, promotions are often made from within the organization. Then, as you move up, you carry along your traditions and ways of thinking. They are ironed in. You fit in the organization. If not, your progress stops. In this form of promotion, quite often there is a built-in roadblock to innovation.

Within business firms, another block to successful innovation occurs when research and development, production, marketing, advertising, and management have difficulty in communicating. Insular groups and cliques form within the company. They resist outside ideas, even from those with whom they are supposed to work closely.

And one final important roadblock is a shortage or lack of necessary facilities. You should have a stenographer when you need her. Typing should not take days to be returned. Necessary equipment should not have to be built and tested over a period of months before you can make the experiment that is indicated. Patent lawyers should not take years before they listen to your proposed specification. If any of these roadblocks are present, then it is frustrating. If they happen over and over again, innovation is certainly slowed down—if not blocked entirely. If you need information on a particular point, you should be able to go to the library. Consultants should be available to help you out. If they are not there when you need them, your ideas and enthusiasms may evaporate.

Roadblocks for management to avoid

The organization is a tool by which the management intends to achieve its goals. It allows management to insure reliable and predictable behavior. However, practices designed to achieve control over the organization may block innovations.

By its attitude and decisions, management can encourage, or discourage, or even stop the flow of ideas upon which it depends for success. For good morale, good decisions are essential. And decisions must be made objectively and based on the best possible information.

On the positive side, management should give the people on whom it depends for innovation freedom of action and time to "think."

Working spaces assigned to innovators should be pleasant, with individual offices to afford adequate privacy. Conference facilities should be available. All these spaces should be quiet and well lighted—but not aseptic.

Management should be careful to reward good performance. It can take several routes in working out suitable rewards:

1. *Prestige.* For example, a better office or a title such as "senior scientist."
2. *Promotion.* Promotion grants more power to the individual, but may stifle his innovative powers.
3. *Money.*
4. *Time.* Permission to take trips worthwhile to the organization, perhaps to attend meetings of learned societies.

To avoid inadvertently putting environmental roadblocks in the way, a manager must understand the nature of all of them.

And, above all, management must minimize distasteful chores. For example, insistance that an inventor prepare the specification for his patent application can cause resentment. To avoid the chore, ideas may be quietly buried in the desk. This is not good for either the individual or the organization. But it is human nature. It has happened.

Furthermore, management must accept and appreciate good ideas. They can always block them by a "killer phrase." But if they kill a good idea, the organization has lost. Yet, in considering new ideas in an appreciative mood, management must always remember that the odds against an idea ever getting into production are high. According to one article in *Industrial Research,* even in successful firms only one idea in sixty ever got into production and paid off. In knowing these odds, management naturally can feel safer in saying "no" than in saying "yes" or even "maybe." But too many "no's" may be disastrous in the long run.

A Closing Remark

It is apparent that a successful innovator must have many unusual personal characteristics in a particular combination. Moreover, any individual doing innovative work must contend with many inside blocks and many outside blocks. Because of the many, many kinds of blocks and possibilities of failure, it is remarkable that any new idea ever gets through the design process. Yet some do! And those that do make our lives better and more worthwhile.

9

Information Processing
by the Brain

Introduction

This book would be incomplete without an example in depth of an application of the systems procedure to a difficult problem. This final example integrates and clarifies many of the ideas presented in earlier chapters.

Unquestionably the most complex information-processing system existing today is the human brain. To construct a meaningful model of this "thinking" system is a challenging and fascinating task. True, models of the brain have been constructed before. In fact, parts of books and whole books have been written about such models. Nevertheless, we believe the attitude and methods of thinking we advocate for a system designer and innovator can define the problems more clearly and take a long step forward in our understanding of what is involved in creating such a model.

One important point stressed throughout this book is that innovation is an information-processing operation. And true innovation is the result of "thinking" by a human brain. Thus, a good understanding of what "thinking" really involves is necessary both for designers and innovators and the managers who work with them. In particular, some human limitations in the processing of information must always be borne in mind. They are of great importance in many man-man and man-machine systems.

Many scientists believe that the ultimate challenge to our understanding is the human brain and how it processes information. The need for such an understanding is pressing; the difficulties are truly enormous. Why?

For one thing, to arrive at a complete understanding, you need an enormous amount of information. This is not the only difficulty. The available information is widely scattered. It is recorded in many, many books and many different technical magazines. Also this information has been dug out by workers in a number of different disciplines.

The total amount of available information is staggering, indeed. Yet, much is still to be discovered. The gaps are wide. And there is another difficulty: semantics. Because each discipline seems to have its own vocabulary, the same (or quite closely the same) idea is expressed by several quite different sets of words. Such a state of affairs makes it difficult to study what is already known.

Physiologists and medical doctors have been observing and studying the human brain for centuries.

The work of histologists who examine the microscopic structure of tissue does not have such a long history. Yet knowledge about the microscopic structure is obviously necessary for a complete understanding of how the brain works.

Medical specialists, concentrating much of their attention on the nervous system and its ills (the neurologists), have made important contributions to our understanding of how the brain receives information and how it uses the results of its processing to control the motor and other systems of the body.

Brain surgery has also contributed. When particular areas are removed, the resulting loss of body functions can be observed. However, successful brain surgery is only a few decades old.

Psychologists study the activities of the brain. They have had another and quite different point of view. They have tested the responses of men (and also animals) to various kinds of stimuli. In this way they have gained a body of knowledge about some of the normal reactions. They have also probed some of the emotional processes with which all of us are more or less familiar.

In most cases the psychiatrists concern themselves with abnormal operations of the brain. Abnormal reactions occur because of trauma or emotional upsets. Some psychologists and psychiatrists who have delved into its mysteries have some very definite ideas about the human brain and its workings.

In the last few decades, a different set of scientists has been exploring particular aspects of the brain. Communications engineers' primary interests lie in the fields of telephony and television. However, they have been and still are measuring the properties of the inputs to the eye and ear and of the signal paths to the brain. They have been studying such problems as the maximum rate of transmitting information through the brain. Among other methods they have used is to have people read a book aloud. We shall examine some of their results later.

Other scholars have tried to make a model network of the neurons in the brain. In some cases they have set up hypotheses as to how the information is transmitted along the nerve pathways in the body: for example, by synaptic transmission.

Along this same line, a group of "cybernetics" specialists have pointed out some particular feedback loops involving the human brain. As a simple example, when you drive a car, you look carefully at the road and the cars around you. Then you adjust your speed and direction accordingly. In other words, you are part of a feedback loop involving the road, the car, your muscles and their responses, and your brain.

One final example: as a result of the enormous impact of the computer on our world, the mathematicians and programmers who write the directions that guide the computer in its work have speculated upon the operation of the brain. There are many references in the literature to a brain-computer analogy.

In the scientific community, other researchers have studied particular aspects. Yet, in many ways, the state of our present knowledge can be likened to that of the three blind men who were asked to examine and describe an elephant. Each felt only a part of the animal. His knowledge was incomplete, and hence his impressions were laughably wrong. Hopefully, our understanding of the brain and its operations is not laughably wrong. Yet, we can be certain that tremendous gaps exist in our knowledge of the brain, of how it works, and about the limits on what it can do.

For quite different reasons, university professors and teachers at all levels in our elementary and high schools need to understand how the brain works.

Why?

Because they have to be able to train those who study with them. If they are unsuccessful, they fail in their job. They are constantly trying to improve their methods. To them, the "information explosion" is no myth. It is here today. The "new mathematics" is an example of their attempt to give more and better training to their students. Obviously, to understand how the brain *works* is a necessity in planning newer and better teaching methods. You don't need to know the details of the physiology. You do need to know how it stores and processes information.

Many nontechnical books are intended to help the layman use his brain more effectively. "Creativity" and "How to Read at Lightning Speed" are just two of the numerous areas covered. You can think of many more. Of particular interest to students and mathematicians are such books as *How to Solve It* by Polya. But the list is long and continues to grow.

Models of the Brain

To help in trying to understand the complexities of the information-processing operations of the human brain, psychologists and electrical engineers have proposed a number of different models to explain some brain processes. These processes have undoubtedly been helpful in advancing our understanding of the problems involved. Some of the proposals have been tested on computers. Computer models of some mental operations have been used to prove mathematical theorems (particularly in geometry). A few computer-oriented experimenters have called their work the forerunner of "artificial intelligence." But even the most optimistic admit that the problem of artificial intelligence is so difficult that they are far from reaching their goal.

This book has outlined a system design procedure. A first step in the system approach is to draw a functional block diagram of the proposed system. One proof of the power of the procedure is to put it to the test: to try it out.

The rest of this chapter presents the system approach to block diagrams representing some important aspects of information processing by the brain. This kind of approach makes clear many of the difficulties encountered in the synthesis of any radically new and complex system.

The approach has a distinct advantage in that it forces careful consideration of how our brains work, and of their inherent and unchangeable limitations.

The results may not be *the final model,* but it is a *possible* model, and contains many elements of truth that will be present when the final model is drawn.

In building our model we shall concentrate on the aspects important for system design. Any attempt to cover all of the possibilities would lead to far too much detail for our present purpose.

The first step in such systems approach is to represent the functions to be performed. This we shall do. Furthermore, we can show the different kinds of blocks required by the physics of the model.

Present-day knowledge cannot tell us how many of the functions are actually implemented. The necessary information is simply not available. Frankly, the gaps in our knowledge are enormous.

A lack of necessary knowledge is a frequent occurrence in system design. The attempt to put together the model helps to pinpoint where further work is essential.

Progress in getting the information depends on research and new measurements. Both "pure" and "applied" research are sorely needed. As an example of the kinds of applied research, take the Bell Telephone Laboratories investigations of the eye, the ear, and the production of speech.

There are excellent reasons for some of the gaps in our knowledge. Physiologists and neurologists are particularly handicapped in making their experiments.

Why?

Because it is extraordinarily difficult if not downright impossible to get into a human brain and make measurements while it is actually alive and working. Furthermore, the working elements—whatever they are—are submicroscopic in size.

As another example of the difficulties of gaining knowledge, take a psychiatrist working with a single patient. To understand his patient's illness may take analysis in depth over a period of several years. And this amount of time is for that one particular person. To analyze another patient requires an equal amount of effort.

Some facts can be deduced from the changes in behavior of individuals who undergo brain surgery. Even then, only the gross aspects appear: not the fine details.

Actually, today we cannot even pinpoint exactly where some basic functions are performed—much less, exactly *how* they are done.

Despite our incomplete knowledge and the uncertainties to which they lead, the model to be developed by the system approach is a step forward. It represents more functions than any other model that we have found in the literature. It explains more kinds of facts. It permits some deductions. And it is an agreement with many ideas expressed by leading psychologists and psychiatrists.

Some of the facts brought out by the model have very practical value. To understand them is important for you as a designer. To understand them is important for you as a management person.

Objectives for the Functional Block Diagram

We want a model to represent the operations of the brain on input information provided by the senses. We want to account for many—but probably cannot account for all—of the necessary operations. Remember we want to represent the necessary functions—and tell how they may actually be executed. However, so little is known about some operations that we simply cannot tell how they are performed.

Some important functions performed by the brain

Just as any good system designer does, we start with a list of functions to be represented by our model.

The brain takes in information. Most of the information important for system design is gained by either of the following methods:

1. Listening to sound waves (such as verbal messages) that reach the brain via the ears.
2. Seeing light waves (such as pictorial or graphic information) that reach the brain via the eyes.

Of course, we have other senses. There is some debate as to exactly how many. But for system design, we can concentrate our attention upon the sensory inputs from our ears and eyes.

The brain can process (*manipulate*) the information furnished it by the sensors.

The brain can operate in two ways: (1) in response to an input; or (2) without a simultaneous input.

Mode 2 is sometimes called "imagination." Complex digital computers can also operate in two modes in some ways quite like those of the human mind. In computer experts' jargon, mode 1 is called "real-time" or "on-line." Mode 2 is often called "off-line": the computer takes in some information, works on it for a while and then puts it out at some appropriate later time.

The brain can store and retrieve information. The information that it stores can be divided into at least four categories:

1. Facts and impressions.
2. Problems that it is asked to solve (or imagines it has to solve).
3. Rules for operating on information (in computer language, programs).
4. Partial results of problems in progress.

The brain can communicate with the outside world. It must react almost immediately to certain inputs when it is operating in the real-time mode. Also it must put out the results of long-time information processing. We communicate information by movements of parts of our bodies caused by contractions of our muscles. Thus the brain may direct the vocal organs and other necessary muscles so that it can communicate ideas verbally. Or it may direct the hands (or, less often, the feet) to display the wanted results by appropriate movements. For example, the movements of the hands may result in a picture or a graph.

The preceding paragraphs give an abbreviated specification for the functional model of the brain. Certainly all of the functions are important. However, they are by no means a complete list of the activities of the brain.

The model which we will work out step-by-step will represent considerably more detail than is expressed by these few basic objectives. It will take into account some necessary relations of the functional blocks to each other. Many of these relations follow from requirements set by the physics of the model. The importance of this statement will become more and more apparent as we develop the details.

Functional Block Diagram: Initial Stages

First block diagram

According to our definition, a system is a set of components arranged to perform a wanted operation or operations. The operation of the system upon its input or inputs produces the outputs. A human being has many inputs and outputs. And you perform many different wanted operations to produce outputs. Thus, you as a human being are a particular example of a system in accordance with the definition.

Because a human being meets the requirements of the definition, it can be represented by Fig. 9-1, which you observed earlier in this book. As the figure shows, a human being has various forms of energy and of objects as both inputs and outputs.

Inputs

A particular input or output may be either wanted or unwanted. Some inputs may be wanted under one set of conditions and unwanted under others. For example, you may enjoy a Thanksgiving dinner. But

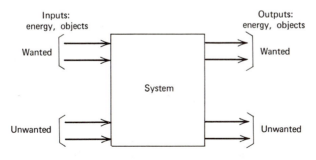

Figure 9-1.

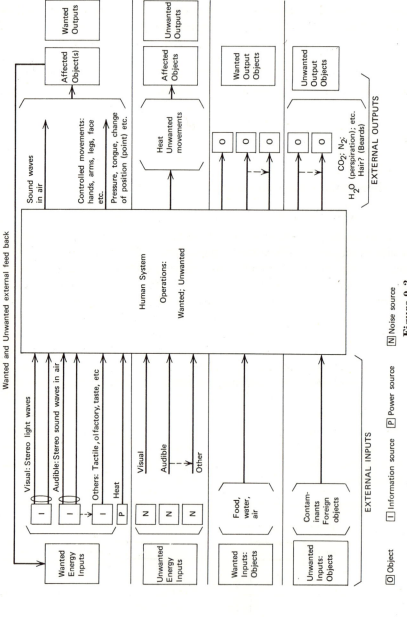

Figure 9-2.

you probably would not be happy about starting a second such dinner just after you have finished enjoying the mince pie and plum pudding of the first.

Fig. 9-2 shows us several kinds of inputs and outputs in more detail.

Wanted inputs: information and energy

The wanted energy inputs in turn can be divided into two categories:

1. Information inputs.
2. Energy inputs.

Time-varying input information energy reaches your brain when it is detected by your body's sensors. For our present purpose, the eyes and ears (your visual and auditory sensors) are the most important. And of these important sensors, the eyes can furnish much more information than the ears.

Under many practical conditions, the sense of feel and the sense of smell are of much less importance to the system designer. By this statement, we do mean that they are never important. You can think of many examples where they must be considered.

We shall give only two examples of wanted energy inputs:

1. Heat energy.
2. Mechanical support.

For you to be comfortable, the heat input from your environment must be kept within a rather narrow range. Of course, if you are suitably clothed you can survive at temperatures many, many degrees below zero—but you won't be comfortable. Also, you can stand desert temperatures well over 100° for a while—but, again, you won't be comfortable.

We ordinarily take the mechanical support for granted. Astronauts who experience the lack of such support under zero-gravity conditions encounter enormous difficulties in performing even simple tasks.

Unwanted inputs: energy

You will recall that an unwanted input is of the same nature as a wanted input. Jamming of a wanted signal by another signal of "noise" is an example of the effect of an unwanted input. An excessive tactile input can cause pain. An unpleasant odor can cause distressing physical results—even nausea.

Excessive input energy can be unpleasant, distracting, or even dangerous. Consider excessive light, for example. You don't look directly at the sun when you are trying to observe its corona. You use a smoked glass (or some equivalent) when you look at a solar eclipse. During a third degree, a suspected criminal may be put under intense light for prolonged periods.

X-rays are routinely used for checking the condition of your teeth and for getting information about possible troubles in the interior of your body. But excessive exposure to such forms of radiation can be dangerous or even fatal. Excessive heat can kill you. Excessive pressure can maim you or cost you your life. Victims of automobile crashes often do not live to testify to the effects of such extreme pressures.

Wanted inputs: objects

As inputs of wanted objects, food (including both nutrients, vitamins, and bulk solids) must certainly be included. Also, the oxygen in the air we breathe; and the water and other fluids that we drink.

Unwanted inputs: objects

Unwanted input objects can be lethal. In the days of the good old Wild West, a bullet from a forty-five was euphemistically called "lead poisoning." The Chinese have used an ingenious form of torture: a slow drop-by-drop of water on the hapless victim.

Unwanted system operations

Notice particularly that Fig. 9-2 also indicates the possibility of unwanted system operations. Disease or drugs (among other causes) can cause unsatisfactory system performance. Such abnormal performance clearly may affect the system outputs. A few important examples will be pointed out later.

Wanted outputs

The wanted outputs that are important for system design involve body movements in response to contraction of muscles. Of course, such movements always require the controlled expenditure of mechanical energy.

Important outputs are those complicated movements that result in our speech: a basic method of communicating with each other. Other methods of communication involve other movements caused by muscles: a lifted eyebrow, a shrugged shoulder, a wave of the hand.

Other movements that you can make control objects: flicking a switch, pressing a brake pedal, turning a steering wheel. Obviously, these are only a few examples. You can think of many more.

Unwanted outputs

Unwanted vocal outputs can be poorly articulated or even irrational speech; stuttering; stammering; or complete absence of speech. Muscular movements may be shaky or palsied; or even absent, as in paralysis.

The unwanted output objects need only be mentioned: the body wastes. Some are solids; some are liquids; some are gases.

Feedback paths

Fig. 9-2 also represents a feedback path from the outputs to the inputs. Sometimes such a path is called "cybernetic" because of its importance in controlling or "steering" the movements of the body. For example, feedback is important in the control of your automobile whether you are driving straight down a superhighway or making a sharp, cornering turn. Using information carried over such feedback paths, the movements of the hand and arm that control the course of the car are adjusted by the brain as it receives signals from your eyes. In this way, you are able to steer the car and follow the course that you want.

Unwanted feedback paths can also occur. One example is an echo when you talk over a telephone line that spans a continent. Unless special devices are added (so-called "echo suppressors"), your voice will be returned to you after traversing twice the length of the line. You hear yourself a considerable fraction of a second after you have uttered a sound. For some people, the returning words or "echoes" can be so disturbing that conversation is very difficult—if not impossible.

To work out a complete model to represent all of the body functions is a formidable task. But let's go back to our stated objectives—to represent the operations of the brain on input information from the senses. More particularly, let's concentrate on those operations of most importance in system design work. We recognize that other operations exist—but we are going to neglect them for our model.

This is a perfectly reasonable assumption to make. But as a result, our functional model need only emphasize the inputs, outputs, and system operations that are important for meeting our objective.

The Modified Model

As we shall see, Fig. 9-3 can represent a model adequate for further discussion. Furthermore, it is considerably simpler than Fig. 9-2.

The simplified model represents only the wanted information inputs: those on which the brain works. Unwanted inputs (such as noise) are omitted because they should be minimized or avoided entirely.

The model represents the wanted energy outputs: those that result from the information processes of the brain. Our attention is concentrated on these outputs because we want to understand the performance of the brain.

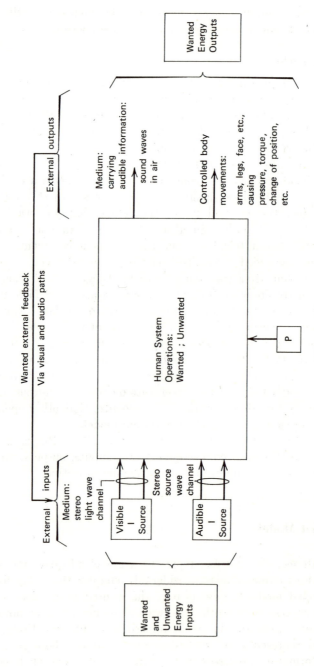

Figure 9-3.

Some further facts become important at this point in the simplification. Of the wanted information inputs, the visual and audible signals are by far the most important. In most cases, tactile, olfactory, and other senses provide far less important information.

Our eyes provide the brain with stereovisual information. Now, the external medium can carry information signals of extremely wide bandwidth to our eyes. In fact, the medium can convey signals over a frequency spectrum far wider than that which the human eye can accept. In the terms of the electronic or filter experts, the information that the mind receives from our eyes is band-limited by our sensors (our eyes). With the right kinds of detectors, very short wavelength (ultra-violet and x-ray) signals can be received; so can longer length (infrared) signals. Our eyes are blind to these wavelengths. Nevertheless, the visual channel is extremely wide compared with the audio channel provided by our ears.

The audio channel also operates stereophonically for people with normal hearing simply because their two ears are separated in space. For many practical purposes, the medium through which we gain our audible information provides a far narrower bandwidth than the visual channel. It can convey vibrations from a small fraction of a hertz to supersonic frequencies. Thus, it is far wider than the capability of the human ear to hear these vibrations.

With respect to information, the medium carrying visual signals to our eyes has far greater capacity than the medium carrying audible signals to our ears. To make the point, the old expression can be paraphrased, "a picture is worth a thousand spoken words."

For our simplified representation, the wanted energy outputs consist of controlled muscle movements. Our vocal outputs are conveyed by sound waves in the atmosphere. Other wanted outputs are due to muscular action. They either affect objects or are observed by the eyes or some instrument.

Both types of movements convey important information that must be taken into account in the model.

In Fig. 9-2, the wanted inputs of objects are largely concerned with supplying the physical needs of the human body. However, all the mental activity of the brain—and, hence all design work—is an information-processing operation. For our model, we shall assume that the brain is supplied with its needs for energy without going into the details as to how this function is actually performed. Thus, we simplify our model and just show a power supply to the information processing blocks. Such a simplification is quite often made in system design work.

A word of caution: such a simplification may not be justified in certain cases. For example, an astronaut in a spaceship must be able to process input information from all of his senses. An unexpected pressure or odor may be very important. Also, he must have an adequate supply of oxygen—otherwise, his brain will function at less than its best level—or even fail completely. However, few people live, work, and process information under such unusual conditions. On this basis, we choose to neglect them. But in doing so, we understand clearly that our model is incomplete.

For similar and obvious reasons, we shall also omit unwanted inputs of objects and wanted and unwanted outputs of objects from our model. The justification is that such inputs and outputs are ordinarily not of primary importance in understanding mental operations. They obviously cannot be neglected under other conditions such as those existing on long space flights.

Also, phenomena involved in unwanted feedback information will be neglected.

After we take into account the simplifications, we can draw the diagram of Fig. 9-3. This will be quite adequate for our present discussion—as we shall see.

Notice that the simplification from Fig. 9-2 to the representation of Fig. 9-3 results from a selection of important factors for a particular study. The selection has been made on a reasonable, logical basis, keeping the objectives in mind.

The Functional Systems

The next step is to begin to fill in more details of our model. We start by breaking it down into a few major functional blocks. For convenience, we shall call these blocks "functional systems," although they might more properly be called the operating subsystems that are found in the human brain. In Fig. 9-4, we represent:

1. A sensory system.
2. A cognitive system. By definition, this system performs the operations that we usually call "thinking."
3. An effective system. By definition, this system performs "non-thinking" operations. These operations can affect the cognitive system and thus indirectly affect the mental processing activities.
4. A motor system that includes among its activities all muscular movements.
5. Power supplies for each system. These supplies furnish the blood that carries the nutrients and oxygen to the systems. As we know,

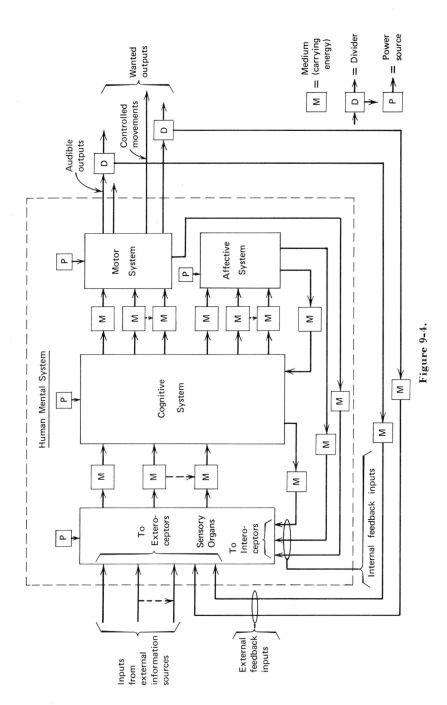

Figure 9-4.

interruption of these power supplies for periods measured in seconds can cause serious brain damage; interruption of a modest number of minutes, death. Power supplies are an essential part of any system. However, in the discussion of the model of the brain they will only be mentioned occasionally and in passing.

Fig. 9-4 also represents many of the necessary and complex kinds of connections between the blocks inside the mental system and also with the outside world.

Remember we are studying the functions of the brain—not necessarily how they are actually carried out in the human body. Much is still unknown about how our brains actually work. Even the actual locations of some brain functions are not pinpointed. At this point in our study, this is not of great concern to us. We know that the functions are necessary and must be represented in our model. Naturally, any uncertainty about how and where a particular operation is performed makes it quite unlikely that we know what kinds of tissue perform it.

Fig. 9-4 also represents:

1. The input media through which the brain receives information via the eyes and ears.

2. The output media that convey the results of the workings of our brain.

Remember that these input and output media are broadband. For example, your eye is sensitive to waves occupying only a trivial portion of the enormous electromagnetic spectrum. Likewise, your ear can hear only a narrow band in the entire acoustic spectrum. Also, the muscles of your mouth cannot create all possible acoustic sounds. The trouble that the Japanese have with the "l" sound of the English language is only one example. English-speaking people also have trouble with some sounds used in German and in French. The movements of your arms and legs can be seen by the eye only over the range of the electromagnetic spectrum to which it is sensitive.

Sensory system

The sensory system takes the various kinds of input signals and converts them into output signals.

Certain input signals come from outside the body; others from inside. External events (either visual or audible) affect the exteroceptors: the eyes and ears. Our two eyes operate as a stereo pair. For this reason, we have a three-dimensional capability when we look at a landscape, for example. When we look down, we can judge how far we are above the ground—"depth perception." Similarly, with our two ears, we are able to tell from what direction a sound is coming.

Some of our sense organs—the "interoceptors"—receive signals from within the body. Such inputs are of great importance in controlling our body movements. Thus, they contribute to our "kinesthetic sense."

The sensory elements provide outputs of coded stimuli (signals) to activate the cognitive system.

Fig. 9-4 indicates the feedback paths from the wanted outputs to the exteroceptor sensory organs. It also indicates feedback paths from the motor system, the effective system, and the cognitive system to the interoceptors.

Cognitive system

The cognitive system receives input signals from sensors over various media are indicated. It may also receive signals from the afferant system:

1. Directly.
2. Via the sensory paths.
3. In both of these ways.

The cognitive system takes its multiple inputs and processes them. Its operations are ordinarily called "thinking." We shall have more to say about these operations later in this chapter.

Some of the outputs (stimuli) from the cognitive system go to the motor system to cause muscle contractions and thereby movements of parts of the body. Other outputs go to the affective system.

The figure indicates the media that form the pathways for these many kinds of input and output signals.

Affective system

The affective system, the "nonthinking system," receives inputs from the cognitive system. After it operates on these inputs, we assume that it generates outputs that go to the sensory organs and perhaps directly to the cognitive system.

Some results of the operations of the affective system can be important when we are "thinking." For example, the outpouring of adrenalin from the adrenal glands when we are excited or fearful can definitely color our reactions. A few more examples will be mentioned later. However, the affective system is not in the mainstream of our study. For this reason, we pay less attention to it than to the other components of the mental system.

Motor system

The motor system receives input signals (stimuli) and uses them to cause and direct mechanical movements of parts of the body. Many (and possibly all) of the inputs come from the cognitive system. However, some may not.

The outputs direct muscular movements that result in speech—by far the most important method of communication between humans. Other outputs control muscles which in turn allow us to operate machines and other devices.

Fig. 9-4 shows internal feedback paths within the body. Information transmitted over these paths enables us to control body movements quite precisely. For example, they direct the muscles so that your hand and arm let you point to a particular letter on a newspaper page.

Others of these feedback pathways carry information that enables us to maintain our balance when we are in an upright position. Many other examples could be cited.

Quite likely, some feedback paths may not involve the cognitive system as it is represented in Fig. 9-4. Some reflex responses occur extremely rapidly. Thus, they may not involve the "higher centers" of the brain.

The functions that depend on internal feedback from the motor system are extremely important for the proper performance of our bodies. But in our present discussion of the model emphasizing "thinking," we shall assume that we are talking about normal people. Hence we shall pay little attention to these feedbacks. We know they are there; we understand their importance; and we assume they are working.

Interconnections and feedback paths

Notice that in Fig. 9-4 there are a set of *forward-acting* connections between:

1. The sensory and cognitive systems.
2. The cognitive and motor systems.
3. The cognitive and affective systems.

There are *backward-acting* internal paths between:

1. The cognitive and sensory systems.
2. The affective and cognitive systems.
3. The affective and sensory systems.
4. The motor and sensory systems.

These pathways are provided by our "nervous system." Some of the details and actual pathways are fairly well understood. Some can only be guessed at or assumed. Some can be inferred by observations of the results of disease or surgery on the operations of the brain.

The coding of the information transmitted over these pathways has been the subject of investigation for many years. According to these experiments, muscular contractions apparently are in response to all-or-none stimulation of individual muscle fibers. A separate signal path to each fiber is required. There are literally thousands of fibers in some muscles. Hence, thousands of microscopic space-separated pathways are contained in the nerve trunks to such muscles. Stimulation of more fibers gives a stronger total contraction of the muscle. But an individual fiber either is contracted or it is not.

Much less is known about the coding over many of the other pathways indicated in Fig. 9-4. This is one of the many gaps in our knowledge.

Fig. 9-4 also represents feedback paths outside the body involving the outputs of the motor system and the sensory organs. Some of our output voice energy is fed back through the external air path to the ears. We are so accustomed to this phenomenon from our earliest childhood that we take it for granted. As we have already said, the medium over which this feedback takes place is far broader than is needed to carry the important frequencies of the voice. Of course, it is easy to measure the voice spectrum as words are spoken. Hence, we know a great deal about the coding around this particular feedback path. We know a great deal—but not all. Scientists are still trying to find out about the differences between people's voices. And when you first hear your voice played back from a tape recorder, you may be quite surprised: it does not sound at all like you imagined. In other words, this feedback path from the vocal cords around to the ears and thence to the brain seems to contain a great deal of distortion.

A second feedback path—and an extremely important one—is via the eyes. They may see the movement of part of your body (say your hands) directly or indirectly via some outside object (for example, a mirror). As previously mentioned, this is the "cybernetic" or "steering" feedback path. It enables us to control some of our body movements with great precision. The eye can see how far off of a target the hand (or other part of the body) is. This "error information" is processed by the cognitive system. It directs the motor system to make an appropriate correction and thereby reduce the error.

Body movements expend mechanical energy. Your eyes are sensitive to light energy. Hence, to measure the "error," an energy change (from mechanical to light energy) is required. In an earlier chapter, you learned that when an energy change is necessary, an energy source must be supplied: in this case, a light source. The light source is modulated by the part of the body under observation to convey the "error information" to the eye. In pitch darkness, when you can't see the path, error correction is quite difficult.

Under some circumstances, the ears can also monitor the position of parts of the body. For example, a guitarist can tell whether his fingers are in the right positions on the frets by the presence or absence of wanted chords. By practice, he trains his "kinesthetic senses" so he knows how to put them where he wants to. If he is sufficiently skilled, he need not check their positions with his eyes.

These examples show some of the ways in which the results of the actions of the motor system are fed to the cognitive system. Such response information undergoes sensory coding. It then becomes input information to the cognitive system along with other inputs from external and internal stimuli.

The next step

Logically, the next step is to show more details of the functional blocks represented in Fig. 9-4. This breakdown might be shown on one drawing. However, a single drawing would have to be large, complex, and very hard to follow. This difficulty arises in trying to represent any complex system. At some stage in the synthesis, another step must be taken.

For this step, you take each functional block and make a separate figure. Each figure must show all of the inputs, outputs, and details of the block that it represents.

We shall now follow this procedure. For our present purpose, it will suffice. But in a very complex system (such as an aircraft, or a spacecraft, or a telephone system) subdivision is continued. In fact, tens of thousands of drawings and figures may be necessary to represent all of the functions in adequate detail and to show all of the necessary interconnections.

Detailed Functional Block Diagrams

Sensory organs

Fig. 9-5 shows more details of the sensory organ block than were represented in Fig. 9-4. In particular, it shows all of the kinds of inputs and outputs.

What does a sensory organ do?

It receives an input signal and converts it to a form which can be transmitted over a signal path and used by the cognitive system. Consider the human eye as an example. Each individual sensory organ in your eye receives its inputs in the form of light energy. It must

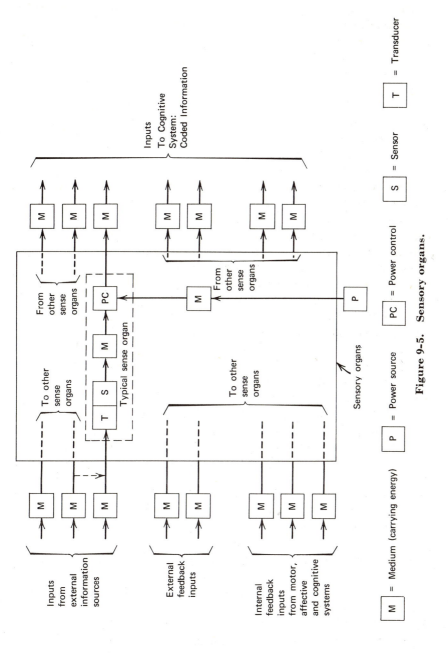

Inputs
To Cognitive
System:
Coded Information

From
other
sense
organs

Typical sense organ

From
other
sense
organs

To other
sense
organs

To other
sense
organs

Sensory organs

Inputs
from external
information
sources

External
feedback
inputs

Internal
feedback
inputs
from motor,
affective
and cognitive
systems

M = Medium (carrying energy)   P = Power source   PC = Power control   S = Sensor   T = Transducer

Figure 9-5. Sensory organs.

convert its input information into a different form of energy that can be transmitted over a nerve pathway to the cognitive system in the brain. Furthermore, the signal energy must be encoded to convey the correct signal information. The encoding consists in changes in quantity of electric energy.

Actually, the energy conversion is carried out in two steps—not in one, as indicated in Fig. 9-5. Light energy (in the form of photons hitting the retina) causes a chemical change. The chemical change causes coded electrical signals to be transmitted over a nerve pathway.

Many details of this two-step energy conversion and encoding are still only imperfectly understood.

We do know the retina of the eye contains three layers of cells: the receptors, the rods and cones; bipolar cells; and ganglion cells. The receptors detect the incoming light. The bipolar cells connect the receptors to the ganglion cells which form the optic nerve that transmits information to the brain. Light passes through the ganglion cells and the transparent bipolar cells before it reaches the receptors.

A rod receptor is an extremely small cylinder: about 1 micrometer in diameter and 50 micrometers long. It is divided into two sections about equal in length. The two are called the inner and outer segments. Light is transmitted down the inner segment to the outer segment where the detection takes place. The outer segment of a cone is the same shape as that of a rod. It is only about 2 wavelengths of visible light in diameter and acts as a sort of waveguide for the light energy. However, the inner segment of a cone is three to six times as broad at the end and has a conical section that gradually tapers to provide a smooth transition to the outer segment.

The rods provide only black and white vision. Color vision depends on the cones. There are about 95 million rods in the retina; only about 5 million cones. Furthermore, there are only about 5 million nerve fibers in the optic nerve. The central 1° region of the retina (called the fovea) contains only cones. In this area, there is about one nerve fiber per receptor. Toward the periphery of the retina the density of rods increases and of cones decreases.

The operating range of the eye is truly astounding. The iris changes the area of the pupil over a range of ten or twelve to one. The eye itself operates over a light range of 10 billion to one; the retina over a range of 1 billion to one; the cone receptors over a range of 1 million to one. Of course, the figures are only approximate.

Despite variations of the light intensity over the field of view and variations of general lighting with time, the colors of objects ordinarily appear to be essentially independent of the light which illuminates them.

In outdoor scenes, the intensity of illumination may vary as much as thirty to one over a scene. Yet a white object in a shadow is normally seen as white; a black object in the sun as black.

When two objects of nearly the same color are put side by side an effect called "contrast enhancement" takes place. Hence, two objects may appear to have the same color when they are separated by some distance. When they are brought close together, they can look very different in color. Because of contrast enhancement, the eye is extremely sensitive to differences in the colors of objects. In fact, when large chips are placed side by side, a normal eye can distinguish at least ten million shades of color. For this to be possible, the responses to color at different points of the retina must be very accurately matched. Because they are closely matched, the color of an object appears to stay the same as attention is shifted to various parts of the object.

The eye can only achieve its wide operating range, constancy of object color, accurate color discrimination, and the uniform field of color perception because of effective feedback controls on its operation. Some experimental work indicates that there are at least two main feedback processes at work: a "time-average" and a "spatial-average" feedback.

Time-average feedback adapts each point in the retina accurately to the time-average of the intensity of light falling upon it. As our gaze continually shifts across the field of view, all points of the retina become adapted to essentially the same average light. Thus they become accurately matched to each other.

Time-average feedback takes care of slow changes of the average lighting condition over a very wide range, but not against rapid changes. To compensate for rapid changes, a spatial-average feedback process varies the response at each point of the retina as a function of a weighted spatial-average of the light received. The spatial-average feedback has a lower range than the time-average feedback. The spatial-average feedback compensates for variations of illumination over the field. Otherwise, a white object in the shade would look black; a black object in the sun, white.

Sensing of changes in light is accomplished by chemical change. A photosensitive pigment changes color when excited by light. The change is called "bleaching." In the rods (the cells responsible for vision in dim light), this pigment is a complex called rhodopsin, consisting of an aldehyde form of vitamin A joined to a large protein called opsin. When it absorbs light, this pigment starts the exitation of the rod. The absorption also splits the rhodopsin molecule into vitamin A and opsin and thereby bleaches the pigment from purplish red to yellow.

One consequence of this method of chemical detection: a deficiency of vitamin A prevents the rods from synthesizing enough rhodopsin. As a result, night blindness in dim light occurs.

Eventually the large protein molecule—the opsin—and a much smaller molecule—the aldehyde of vitamin A—recombine and the original pigment molecule is reconstructed.

The bleaching rate is proportional to the light energy, and the regeneration rate proportional to the number of bleached molecules present. When a constant light is turned on, bleaching begins and the concentration of bleached molecules rises. In complete darkness, there are no bleached molecules. However, with more bleached molecules, more tend to regenerate. Thus an equilibrium condition is reached when the regeneration rate reaches the bleaching rate.

Both bleaching and regeneration require appreciable time. If the average light changes at a slow rate, the eye can perceive it as flicker. On the other hand, if the rate exceeds twenty to thirty times per second, the eye does not follow the abrupt changes. This fundamental property of the eye is utilized in the projection of motion pictures. Too slow a rate causes a "flicker." Rates above those which permit the observation of flicker give the impression of a steady illumination. The eye sees only changes in the field of view.

To complete the picture, we must assume some sort of selective mechanism operating upon the eye. For example, with a little experience you can use a monocular microscope without closing the other eye. Some workers studying the mechanism of the eye have suggested that the cones are scanned by some process. By this scanning, all the cone photo-pigment molecules in the retina at a given value of outer segment radius are sampled at the same instant. Furthermore, a particular molecule would conduct only for a short interval when the scan reaches it. The scan would cause the output from the cone to vary with time. Thus the output would be such that both the color and black and white information would be obtained by a demodulation process in the cognitive system. The model which we are discussing will put such a selection process in the cognitive system.

Present evidence indicates strongly that neither the retinal receptors nor the bipolar cells generate impulses. Some earlier work assumed that impulse was a basis for all neuro-electrical activity. The neuron acted like an ON-OFF switch that fired "all-or-none." As far as the optical signals are concerned, this concept may be false. Nevertheless, the exact method of operation is by no means thoroughly understood today. With 100 million light sensors in the retina and 5 million neuro pathways to the brain, it is easy to see the difficult problem confronting physiologists attempting by experiment to find out exactly what is going on.

Just as for vision, hearing involves an energy conversion and a coding.

The input energy is in the form of sound waves in air—in other words, acoustical energy. The evidence indicates that the energy transmitted to the brain is electrical.

Just how the energy conversion and coding are performed is not known with great certainty. The mechanical operation of the ear is fairly well understood; the neural operation is not.

Sound waves enter the outer ear (the pinna), travel down the external canal, and cause vibrations of the ear drum (tympanic membrane). The mechanical vibrations are conveyed through a chain of middle-ear bones to the air-filled middle-ear cavity. Then they go to a liquid-filled spiral (the cochlea) contained within the skull's temporal bone. One of the middle-ear bones—the stapes—communicates the vibrations to the cochlea liquid.

The cochlea is divided into upper and lower compartments by a partition. The largest part of the partition is the thin basilar membrane.

For all incoming sound vibrations, waves start at the middle-ear end of the basilar membrane and travel slowly to the other end. It takes about 5 milliseconds for a wave to travel the length of the membrane (about 35 millimeters). The place along the length where maximum amplitude is reached depends on the sound wave frequency. Each frequency has its own particular place. Thus, the membrane spatially separates a complex sound into its vibrating components.

The vibratory motion must be converted into a form so that appropriate signals can be sent via the nerve pathways to the brain. The conversion is accomplished by about 30,000 "hair cells" in four rows along the basilar membrane. A tuft of tiny hairs from the top of each hair cell receives the vibrations. The base of the hair cell contacts the acoustic nerve fibers to the brain.

So much is known with a fair degree of certainty. But the exact way the signals to the brain are generated and coded is still the subject of speculation. True, there are guesses about what goes on, but they are only guesses. Another gap in our knowledge!

The motor system

Fig. 9-6 shows all of the kinds of inputs to the motor system. It also shows the outputs causing the muscular movements in which we are interested. The motor system functions to move the bones and tissues of the body. In particular, it effects the movements of the parts of the body involved in speech. As a result of these movements, sound waves in air are generated. It also causes the movements of our arms, our hands and of our legs and feet. By them, we can perform such operations as controlling machines, pointing our fingers, writing, and drawing pictures.

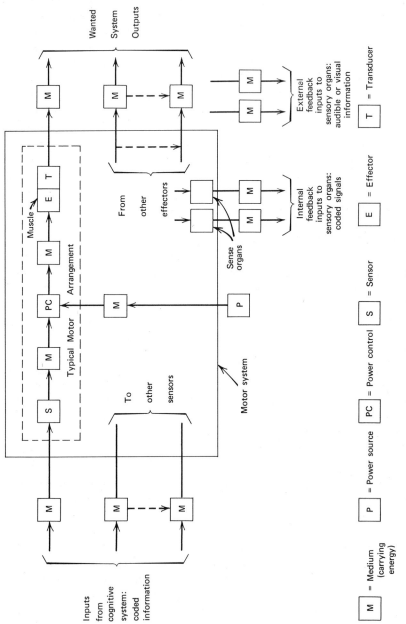

Figure 9-6. Motor system.

The motor outputs are in the form of mechanical energy. The inputs to the muscles are in the form of electrical energy. These inputs perform at least three functions: (1) they indicate the movement to be made; (2) they control the direction (or directions) of movements in space; and (3) they control the amplitude of movements. All three functions may change rapidly with time and, particularly, in rapid speech.

Since the input signal energy and the output mechanical energy are of different forms, a source of mechanical energy is necessary in the motor system. This is shown in Fig. 9-6. Also required is a control of this mechanical energy by the input electrical signals. In many parts of the body, a transducer (such as a lever) changes the mechanical effect of the muscle contraction which causes the movement. Obvious examples of levers are found in the arms and legs, but there are many others in the trunk and extremities.

Affective system

For our present purposes, the affective system seems much less important than the other blocks shown in Fig. 9-4. For this reason, it is convenient to show it as an appendage to the cognitive system in Fig. 9-7. In this figure, the inputs to the affective system come from the cognitive system. The outputs go to the cognitive system and also to the sensory organs: they form internal feedback paths in the body.

According to the definition given earlier, the affective system performs the "nonthinking" operations. But remember that certain mental phenomena (such as anxiety) can affect both the cognitive and affective systems. Hence, they can affect many body process via the feedback paths to the cognitive system and to the sensory organs.

The cognitive system

We now come to the most important block of all: the cognitive system. It sifts all of the sensory inputs and selects from among them the items of information that because of vividness, interest, or usefulness are to be acted upon. Many of these processes are believed to be carried out by the cortex of the human brain.

The cognitive system furnishes the inputs that set in motion the effective and motor systems. Notice particularly that the cognitive system is the master: its outputs activate the appropriate muscles, glands, and other parts of the body.

As already pointed out, the brain is an extremely complex organ. It is estimated to contain 10 billion nerve cells or neurons sheathed by 100 million glial cells. And these figures do not include the connecting maze of the peripheral nervous system going to the muscles, glands.

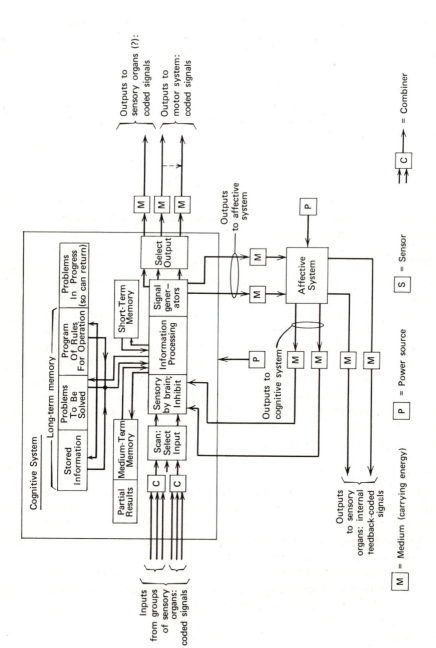

Figure 9-7. Cognitive and affective systems.

and other parts of the body. These numbers are far greater than the numbers of components and interconnections of any computer existing today—or likely to exist for many years in the future.

Fig. 9-7 shows the inputs from groups of sensory organs and outputs to the motor system. As explained earlier, these inputs and outputs are in the form of coded electrical signals.

Fig. 9-7 also indicates outputs to the sensory organs. These outputs are only assumed to exist. However, without assuming their existence, it is difficult to understand how certain observed phenomena could take place.

Outputs are also indicated to the affective system. Here, too, the outputs are assumed—not actually identified as such. But we all can understand that the adrenal gland pours forth adrenalin when we are frightened (for example, by an angry bull). We also are aware that when we are anxious or under pressure we perspire, we blush, we tremble.

In Fig. 9-7 the input signals to the cognitive system (for example, those from the eyes and ears) are acted upon by a combiner. Psychologists sometimes call this operation "integration." In this way, separate input signals are combined into a composite signal for processing. Thus, we do not see the individual outputs of the sensory elements in our eyes. Instead, depending on what we are looking at, we may see a distant mountain or our cat lying at our feet. The necessity for this function seems obvious from these two simple examples. Hence, it is assumed to exist—but not much is known about the actual mechanism by which it is accomplished.

Three additional operations may also be performed on the input signals: selection, filtering, and inhibition. Since these may be thought of as different forms of selection, only a single block is indicated on Fig. 9-7.

We have already mentioned some scanning mechanism that operates on the retina output under control of the cognitive system. We set up the necessary requirements for a selection process in an earlier chapter. Thus, the cognitive system must provide the address of the information it wants. For example, to view an object in space, we may turn our head, or move our eyes, or both. Such operations are "second nature": we do them without "thinking." Yet, actually, they are extremely complex. And that is not all. You can select a particular area or object in a large scene and concentrate your attention on it. Girl-watching at the beach is one example. Or you may choose to observe a humming bird flitting from flower to flower. In doing so, you neglect the myriad of other details in the scene in front of you. When you drive a car on a superhighway, you concentrate on the road—not on the scenery. Or else! These are all examples of selection of the visual images in which you are interested and disregarding those in which you are not.

You can perform a somewhat similar operation on sound inputs. You can disregard a large amount of unwanted noise and concentrate upon the sounds you want to hear. A mother can be talking with a large group of people and yet hear the cry of her baby in the next room—even though the door is closed. But you can do even more. For example, at a concert you may choose to concentrate on a particular section of the orchestra—say the violins. Or you may concentrate on a single piccolo piping away while the kettle drums are making a fearful racket. In doing so, you are unconsciously selecting the high-pitched "treble" notes and disregarding the "bass." Thus, our ears and brain together can allow us to disregard noise, to listen for particular notes, or to locate the source of sound in space. Just how we do all these things is not well understood. But it is under careful study in numerous laboratories despite the fact that such phenomena are half-jokingly called the "cocktail effect." Presumably, this phrase is apt because you can carry on a long conversation with a friend—or even an ac-quaintance—in the midst of the din at a lively cocktail party. The fact that a hundred other people are also trying to carry on "intelligent" conversations at the same time does not prevent your establishing the necessary rapport with your companion of the moment.

The existence of "inhibition" in the central nervous system is unques-tioned but the details of how it operates are very obscure. In a sense, inhibition may be thought of as a special capacity of the cognitive system to neglect unwanted input signals. One example has already been given: unwanted noise at a cocktail party. In some way, you just don't "hear" the noise. Take another example. You can look through a monocular microscope with your right eye. With a little practice, you can completely neglect the signals coming from your left eye. You do not "see" what the left eye is looking at. An Indian fire-walker is said to feel no pain when he treads over red-hot coals. Why doesn't he? Physicians and metaphysicians have speculated. But nobody knows the real answer.

All the information gathered by the sensors of the body and perhaps information from the affective system flows over the pathways provided by our nerves to our brain. In the brain, all of these signals are pro-cessed to generate the outputs that direct the muscles and organs of our bodies.

The brain must generate appropriate output signals to direct a particu-lar part of the motor or of the affective system. To be able to do so, it must select the output pathways over which the signals generated by it are to pass. For example, to direct the hand to draw a picture, signals must be transmitted to the muscles of the hand and arm to control the wanted movements. Thus, two separate kinds of operations are necessary: (1) generation of the proper signals; and (2) selection of the desired paths. These blocks are indicated in Fig. 9-7. Remember that selection always requires some form of address information so that the proper pathways—and only the proper pathways—are permitted to carry signals.

Information-processing operations

We now come to the most fascinating subject of all: the information-processing operations. For it is in these operations that you take in information and remember it. You can process input information in many different ways and generate outputs resulting from the work.

It is believed that the processing operations are carried on by the cortex of our brains. But many mysteries about the how and where are yet to be explored. However, since we are interested in understanding some of the functions and not the details of how they are performed, we set ourselves a much easier (but necessary) task.

One of the remarkable things about the brain is that it can be activated in two quite different ways: (1) by a sensory input; or (2) without a sensory input. In the first case, the input may be said to be "perceived." In case two, the operation of the mind may be called "imagination." In this context, the presence or absence of a sensory input may be said to determine whether something is "really" present or only imagined.

If you are used to talking about computers, you might say that case 1 is somewhat similar to a computer working "on-line"; case 2, to "off-line" operation.

Psychologists sometimes assume that the sensory signals are *usually* able to override the imagination. Usually—but not always. Hallucination can occur under a relatively rare combination of conditions: a strong imagination and no corresponding sensory input. In hallucination, the image is so strong that it is believed to be real.

We can look at these facts in yet another way. Thus we can say that the brain can process information provided by input signals *and* information stored in its memory: case 1. Or it may use only information stored in its memory, that is, imagination: case 2.

The four levels of memory

In our everyday discussions, we say "my memory" or "I remember." Everyone knows—or thinks he knows—what we mean when we use such expressions. We store in our brain an item or items of information. We may or may not ever want to retrieve what we have stored.

Actually, the facts about memory are much more complex. There are at least four kinds of *functional* memory or storage. These are:

1. *Built-in*. There is evidence that we possess some built-in knowledge when we are born. Such an assumption can account for such observable facts as the "suckling" instinct of a new-born baby. Cats, dogs, and other animals seem to have more in-born knowledge than human babies do.

2. *Short-term memory*. This memory level stores information for a few seconds or so. For example, when you look up a telephone number, hopefully, you remember it long enough to dial. Some of us can't even remember it that long. In fact, most of the visual, audible, and other input signals that you receive during your waking hours are promptly forgotten. Probably, it is better so—otherwise, our memories would be cluttered up with even more useless information than they now are.

3. *Medium-length memory*. This memory level retains information for a few minutes to a few hours. As a student, you can use this level of your memory while you cram for an examination. After that confrontation is over, you can forget most of what you stuffed in. But this level has other uses also. For example, if you are working out a long problem, it will hold intermediate results for you until you need them later. Also, this ability to hold partial answers allows you to start and stop your thinking. Otherwise, a telephone call would be even more annoying than some of them are anyway. Your medium-term memory enables you to pick up where you left off.

4. *Long-term memory*. Long-term memory is for the storage of information you don't want to forget. These may be isolated facts such as important events in history and when they occurred. They may be the rules of arithmetic that enable you to add a column of figures. The word "program" is often used by computer people for a set of rules for solving a particular mathematical or other problem. If you have to solve problems—and you are wise—you will store in your memory the successful and unsuccessful methods you have used in solving various types of problems.

By definition, in (1) the information is built in. You do not have to put it there—in computer language, you don't have to "write" it in. In (2), (3), and (4), you do. Consciously or unconsciously, you have to write it in. In all four cases, you may want to recall the stored information: that is, to "read it out." At least in the last three cases, you may wish to forget information that you have stored or "erase" it. Perhaps you have learned it is useless, wrong, or unpleasant. In fact, the phrase "forget it" is part of our language. How do we erase? A good question—without a good answer.

The need to retrieve information you have stored in your brain cells raises some extremely important questions. For example, to retrieve information, you have to find it. To find it, you have to know where you put it. Unfortunately we know little, if anything, about how we actually determine where a particular item of information has been put in the memory part of our brain.

We do not know how the information is stored or coded. Some recent experimental work leads some investigators to think that information is coded by changes in the protein molecules in our brains. But where? And how?

By experiments made on animals and men, we do know that the storage of information takes time—time measured in fractions of a second. But again we do not understand the basic phenomena that must be involved.

How do we retrieve information? We know little about this process either. But some search techniques can be ruled out. For example, we are all familiar with almost "instant recall." You are asked a question. Your answer is almost instantaneous. It is almost inconceivable that you could have searched the billions of memory cells in your brain in the time that it takes for you to answer. So one-after-another, or sequential searching, seems ruled out: it would be far, far too slow.

Therefore, some other—and more rapid—process must be the key. In some way, when an item of information is stored, it must carry some kind of a tag or address. Then if need arises, it might be retrieved by going directly to that address—and to no other. One bit of evidence that such a method actually occurs is the use of "association" in memory courses. You associate a code word with a particular card in a deck that you are shown. By recalling a sequence of these code words, you can recall the sequence of the cards in the deck. Possibly you learn the multiplication tables the same way. But the assumption of regaining information by going to an address raises another troublesome question. How does the brain do the addressing? Where and how does it remember the address? Certainly not in our conscious memory. But how? Is there only one address for a particular item of information? Or more than one? We don't know.

There is another possibility. Possibly, our brains retrieve information as we look up a telephone number. If you look up the number of your friend John Smith, you don't start at the front of the book with the A's and go all the way through to the Sm's. You take a try at about where you think the S's are in the book. If you are far off, you use that information to go forward or backward in the book to get nearer to the page on which your friend's name is located. Several trys may be necessary, but finally you get to the right page. Then, and only then, do you run your finger down the page until you find the name you want. This is a form of random access (as the term is used on computer work) followed by sequential access. You first get in the right ball park, and then you go around the bases until you are home.

There are intriguing ideas about how the brain functions. But they are only speculations. Nobody really knows. However, the functional approach points out many requirements. These must be met by an assemblage of building blocks working together.

There are other mysteries about the operations of the brain. For example, many psychologists hold that a repeated sensory input in some way is directed to the same place in our memory. Therefore, repetition reinforces the stored information in some way. If this is true, then the brain must indeed have a rapid addressing facility so that it can direct the information to the proper location. Otherwise, you have to assume that information is stored temporarily and redirected at a later time. Such a method of operation might be called "single location in memory." Your telephone book works this way. Another possibility is to enter the new information in the storage file of your brain when it is received. The same item might be filed in many places. This would be extremely wasteful of memory cells—but have you ever heard of any limitations on the memory capacity of the human brain? "Multiple location" storage of a piece of information could have system advantages. Recall might be facilitated if a particular piece of information could be found at a number of different addresses. Presumably, it might be recalled by more than one associated idea.

Frankly, we don't know how our information recall works. And we shall not understand how our brains work until we know the answer.

Does our memory of information deteriorate with time? Possibly so. And possibly not. Take something that happened in your childhood, for example. You had little need for this information for many, many years. Hence, there are few (if any) recent associations. Yet, as we all know, under the proper circumstances, you can dredge things up from your "subconscious." Sometimes we say they have been stored in the inner recesses of our brains. Yet, they can be recalled in great detail. This has led some psychologists and psychiatrists to wonder if we ever forget anything. In fact, a well-known technique in psychiatry probes these forgotten or blocked-off memories. It is based on the belief that such suppressed thoughts may cause abnormal behavior in some cases. If these ideas are correct, they furnish another amazing example of the uniqueness of the human brain because other forms of memory all deteriorate with time. Some forms deteriorate only slowly—for example, letters written on stone. But even they fade—and sometimes become illegible—within the span of a human lifetime. Possibly, the memories of the mind are more enduring than those preserved in stone. But who knows?

Information processing

The powers of the human brain to process information are great, indeed. However, a relatively modest number of different kinds of functions can be performed. Some of these will now be discussed.

You can be taught how to compute. For example, you learned to solve arithmetical problems: to add, to subtract, to multiply, and to divide. The rules (or programs) for performing these operations are stored in your brain. When you need to multiply two numbers, you retrieve the multiplication tables that are necessary from your memory. As you go on in your study of mathematics, you learn how to solve trigonometrical problems, differential calculus problems, integral calculus problems, and so on. You solve them by taking the input data to be worked upon and applying the rules which you have learned and stored in your memory.

You can also solve logical problems. You can deduce conclusions from a given set of input information. In doing so, you apply rules of logic stored in your memory.

But you can also reason from the particular to the general: you can draw conclusions by induction. Computers can be programmed to perform deductive operations; but, as yet, it is doubtful if they can perform inductive operations. In inductive problems, the human brain can still far surpass the most powerful computer ever built.

You already know that it is possible to discriminate between two colors with a remarkable precision. Of course, this process of discrimination is performed by information processing. You can also listen to two audible tones and tell which is louder, or higher in pitch. You can tell whether one object is larger than or equal to another object in size. Physiologists have made many experiments upon our ability to discriminate between almost similar things. Such experiments are possible despite the fact that we do not know exactly how the brain makes the discrimination.

Even more mysterious is the ability of the brain to compare a particular object or phenomenon with some reference contained in its memory. Have you ever said: "That's the most beautiful rainbow I've ever seen?" When you make such a statement, you are unconsciously comparing the rainbow you see with all stored memories of rainbows that you have seen before. We make such comparisons all of the time. This is the hardest, or the easiest, or the longest, or the shortest—we make such comparisons every day.

We all use our brains when we make decisions. We decide that we don't like so and so, or that we're going to the mountains for a vacation instead of the seashore, or that we will buy a Chevrolet instead of a Ford. Such decisions are choices between alternatives. If they are "rational" decisions, then the alternatives have been evaluated objectively, the rules of logic stored in our memory have been applied, and decision is made. All of this may occur without our ever knowing that it is taking place. In fact, in most of our everyday decisions, we never give a thought to the decision-making process. We decide to cross the street now, instead of waiting. Only if the consequences or a wrong decision might be unpleasant or expensive, do we ordinarily go through a step-by-step process of making a decision.

We can take in information and change its order. As a simple and almost trivial example, we can write words horizontally or vertically. We can rewrite the words in sentences and the sentences in paragraphs. In other words, we can *edit* our thoughts. In editing, we may strive for extreme brevity and compression. On the other hand, we may add adjectives and adverbs to try to evoke visual images by our writing. In the language of information theory, such changes are called changes in the "redundancy." The information content remains substantially the same, but the word count may differ considerably before and after the editing.

As another example, consider the ordering of WANTS and DON'T WANTS in setting goals and objectives. In this case, the ordering depends on a set of abstract subjective preferences—not on spelling or grammar.

If we are trained, we can change the language of our thoughts—say from English to French. All of us do not have this ability, but many human beings can translate between two or even more languages. A somewhat similar ability is exhibited by telegraphers who can transmit and receive in two or more telegraphic codes.

Many of us are particularly good at pattern recognition. This ability thus far has eluded the best of the computers. For example, you can read a book whether it is printed in any one of a hundred or more different fonts. It does not bother you particularly if the letters are slanted, or are large or small, or even mutilated. You can read the handwriting of many different people. Some of it is almost illegible—to be charitable. Not a great deal is known about how we are able to do these pattern recognition processes. They are truly remarkable compared with those of any existing machine.

As another example of the abilities of the brain, it can pick out similar objects among a large group of apparently random shapes. You can glance at a figure that is not quite complete and fill in the details by retrieving from your memory information about how such details ought to look. You can recognize a caricature of FDR or Churchill at a glance. You read by context—not by consciously noting every letter—or even every word. And yet you are able to extract important information as you read extremely rapidly.

You are also able to recognize audible patterns: you know with whom you are talking over the telephone even though you may not have heard the voice for months or even years.

Experts are only beginning to understand some of the peculiarities of voice patterns. Yet you can recognize the patterns and store them in your memory so that you know them when you hear them again.

Finally, in your pattern recognition, you can combine visual and audible inputs. You can understand changes of meaning conveyed to you by voice tones, facial expressions, movements of the hands or feet, even by shifts of the eyes or blushing. Again, we do not know how you are able to do these things. We only know that you can. And we also know that no machine today has such capabilities.

As pointed out earlier, you can process information by using stored impressions. Among other kinds, these impressions may be visual or auditory. They are the result of previous sensory inputs. When you process such information without immediate sensory inputs, you use your imagination. As far as we know, imagination is a purely human mental activity.

Our brains possess another important attribute: the ability to continue thinking about an idea until some new sensory input takes over. Psychologists sometimes call this process *reverberation* of the signals in the cognitive system network. Laymen may use the word *concentration*. We all possess this ability to a very considerable extent—even though some of us may be called "scatter-brained" at times. Children can concentrate briefly, but as you get older this ability improves.

Some psychologists believe that attention shifts automatically to the strongest input of the moment. Thus, a stronger input causes a change in the focus of our conscious attention. In their technical language, they sometimes say *hyperfacilitation* functions to amplify the strongest input among those momentarily present. Exactly what processes take place in changing the focus of our conscious attention is not entirely clear. Yet we can and do change our attention from time to time.

Finally, we can work on a problem for awhile, be interrupted by another problem, solve the second problem, and then return to the first. In other words, we can say that we can perform *multiprocessing*—work on several problems in sequence, turning from one to the other from time to time. This is indeed an amazing ability. It would seem to require some form of temporary storage where you can put partial results of your mental efforts until you have the need or opportunity to return to them.

How the parts of the cognitive system to perform these numerous and complex processes are distributed in our brains (and perhaps elsewhere) is not completely known. It seems unlikely that the "reflex" response when your hand touches a hot stove is the result of information processing in the higher levels of the brain. Thus some processing may not need to go to the higher levels. Some responses may be "instinctive."

In conclusion, let us mention another example: When you look at an object, is your information processor retrieving information from storage as you look? In other words, are viewing and information retrieval simultaneous or overlapping? Possibly, you view the object, then decide that certain information must be retrieved and then retrieve it: sequential operation. Conceivably, your brain could operate either way, or even simultaneously at one time and sequentially at another. For at least some cases, strong arguments can be brought up to make one believe that the operation is sequential. For instance, take a simultaneous translator listening to a Russian speaking. He then conveys the meanings of what was said in idiomatic English. Clearly, he must first hear what the Russian says, search his memory for the exact English equivalent, and then speak. Thus, he acts sequentially. If he tried to overlap, the task would be impossible. He would have to try to guess what the Russian was going to say next. Clairvoyance would be necessary.

Examples of data processing by the cognitive system

We have already mentioned many of the information-processing activities of the cognitive system. For example, you take in input information into your brain and manipulate it. You change its order, language, etc. You compute.

You perform logical operations. In particular, you can draw conclusions by going from the general to the particular. Also, you can employ induction and go from the particular to the general. You routinely solve many kinds of problems by logical reasoning or computations or both.

If patterns exist, you often can recognize them in visual and in audible inputs. You can sort input information or objects. For example, you can take a deck of cards and arrange it with each suit in order from ace to king. You can sort your monthly checks so that those to the grocer, to the bank, and to the doctor are together. In fact, you have to do so to work out your income tax.

You can also collect or collate information or objects from different sources. Ordinarily, a new book (such as this one) contains information gathered from many sources. The information is selected, arranged and rearranged, added to and thought about, until finally the draft is ready to edit.

These are some of the many operations of the human brain. But they are by no means all. Some other important aspects will now be mentioned.

Learning

Earlier, it was pointed out that there are certain "built-in" connections in the brain. These connections represent constraints on the brain capacity. Our bones, muscles, and ligaments constrain our bodily movements. For example, you cannot turn your head more than about 90° to the right or left. Why? Because of the built-in constraints imposed by bone shape and the ligament and muscle limitations. The range of movement of your elbow joint and knee joint are also constrained.

It seems only reasonable to assume that the brain organization also has constraints. For example, the interconnections of the brain cells may impose some operating constraints—but we don't know. Some psychologists believe that when you are born the storage cells of the brain are empty. If so, then we can invoke the "organization" of the brain to explain the instincts with which we and other animals are born. Otherwise, we can assume that in the beginning these instincts are contained in a few of the storage cells. Or, perhaps they are present in both places. We just don't know.

Whatever the explanation for our instincts, in the beginning all is not complete chaos. Some organization exists. Furthermore, as time goes on, you gradually fill the storage cells of your brain with your own special contents based on your experiences.

"Learning" implies setting up relatively permanent alterations in the state of your long-term storage cells. These alterations are made by your personal experiences. Your experiences are unique with you. They are unlike the experiences of any other human being.

The ability to recall an item of information means that sometime in the past you were furnished that information by your senses, or else it was in-born.

If the particular item has been furnished many times and particularly if it has been vividly experienced, some psychologists say that information has been *reinforced*. Reinforcement is believed to make that item of information easier to recall. Note that we have carefully avoided saying whether reinforcement increases the strength of the information stored in a single location or the storage of the information in more than one location.

Repeated recall of a particular item of information seems to make it easier to remember. Is the opposite true? Does stored information fade with disuse? Some psychologists assume that weakening does take place—but very slowly. Thus, at least in theory, after many years, the information is lost. A lengthy period when the information is not required permits deterioration until the storage location is no longer functional.

Perhaps deterioration does take place. But if it does, it must be subject to some kind of a selective process. Otherwise, how can you possibly explain the ability of a senior citizen to recall events of his childhood and younger years in great detail? He remembers them vividly—even although he may not have thought about them for decades.

Other aspects of learning are also both interesting and pertinent. For example, when we look at a distant scene, there is an image on the retinas of our eyes. There is a vast difference between the richness of what we can perceive by using the information processes of the cognitive system and a momentary retinal image. For example, we possess the property of space perception. Thus, by the operation of our brain we can be conscious of one more dimension than that possessed by the retinal image. Furthermore, an object tends to be the same size despite changes in the viewing distance—even though the size of the retinal image changes. This is called "spatial constancy" possessed by our cognitive systems.

We have to learn these abilities. They are not present at birth. However, in a normal child, they are learned at a very early age—perhaps even in infancy. But how we learn them is not well understood. The Gestalt theories in psychology have been formulated to try to throw light upon how these abilities are learned. Some recent evidence seems to show that a very young baby can register the required sensory information. However, he cannot process it. In other words, the program for processing such information must be learned. This is an interesting point. It brings out again the two separate parts and functions of the brain which we have used in our building block approach.

Psychologists have made many experiments directed toward gaining an understanding of how animals and human beings learn. One way in which some of their experiments are conducted is to reward the subject of an experiment for making a correct choice and punishing him if he does not. He "learns" to make the choices desired by the experimenter because when he does so he is rewarded. Wrong choices are punished.

Rewards and punishments often—if not always—cause some emotional responses. In some way, these emotional responses have consequences that affect the thinking. When stated in this way, it is easy to see that the reward and punishment technique is not quite straightforward. It affects the process of learning via the emotions. This is an interesting fact, and is not intended as a criticism of the technique.

Role of the subconscious

There is an old saying that you can solve with your "subconscious" mind any problem that you can solve consciously—and some that you cannot. The "subconscious" processes information below the conscious level, but the results can be called up to consciousness by a proper stimulus. All of us use our "subconscious." Some of us are able to do so more effectively than others. For example, before going to sleep, you may go over and over a problem. You fix it in your brain. Then you go off to sleep. Next morning, perhaps while you are shaving, the answer comes forth in a flash.

What has taken place?

Clearly, while you were asleep you were searching the storehouse of your memory for information pertinent to the problem. Furthermore, you must have been processing and arranging the information—perhaps trying out many arrangements. Your eyes were closed, you were not listening to outside sounds. You were asleep. Therefore, the answer to your problem must have been found by your cognitive system. But it was not stored in your memory. In particular, you had never encountered this problem before. Because it was new, *you had to retrieve and process information.* The necessary information may have been long-forgotten. It may have been a fact. It may have been a procedure. It may have been both. But you had to process information to get the answer.

Imagination and innovation

You can use your "subconscious" while you are awake as well as while you are asleep. You can create images without perceiving the objects. In fact, you can day-dream—and all of us do sometimes.

The "thinking" we do in imagining new arrangements of facts, new arrangements of objects, new ways to do processes—such thinking is the basis of innovation. And it is innovation that gives us our new books, our new paintings, and our new inventions. Without imagination, it is impossible to create the new. And almost always, the creation of something new is a result of the workings of the subconscious mind.

"Flash of genius"

There is one remarkable aspect of the solution of a problem by the subconscious. The answer comes suddenly—almost in an instant. You don't know how you got the answer. You just know that it is right. It may take you hours—or even days—to try and raise to the conscious level the actual steps that led to the answer.

Why does the answer often appear suddenly?

Possibly, you hit upon the precise association and hence upon the right address to reach a needed piece of information stored in your brain. Armed with the right address, recall seemingly can be relatively rapid. With the necessary information at hand, processing may be speedy. The information that you need may be a fact. It may be a new program. Or even just an ordering of steps in a chain of reasoning that leads to the answer. When all of the essential information is gathered together, then the information processing function of your mind arranges them in order and gets the answer. Eureka! Another stroke of genius!

To sum up, the "flash of genius" seems to be a perfectly reasonable phenomenon in terms of the model put together by system thinking.

Power of suggestion

Everyone can be influenced by suggestions. Some people are more easily influenced than others. Nevertheless, do not underestimate the power of suggestion.

The television show "Candid Camera" ran a revealing experiment some years ago that showed how suggestible most people are. For their experiment, they filled a number of perfume vials with plain water. They then offered them to a series of adults with the statement that the vials contained a perfume. The subjects were asked to tell the first thing that the odor of the vial called to their mind. The answers were varied. Of course, since the program was "Candid Camera," some were very funny. Among them: jasmine, sandalwood, outdoorsy, a woman, onions, spring, a hospital (alcohol), very delightful, soothing, spicy, almonds or cocoanut, and cinnamon. Thus there were a dozen different thoughts evoked by the vial of water masquerading as perfume. Only one person commented on the "lack of strength."

Such is the power of suggestion.

The power is so great that it should be used in design with care. But it should be used to promote right—not wrong—answers. If it is used to promote wrong answers, it can cause the failure of a project.

Inhibition

The power of suggestion is one of many factors that can inhibit thinking. The existence of inhibition in the cognitive system is unquestioned. But the details of how it operates are obscure.

We have already mentioned inhibition of the sensory system. This can be expressed in terms of the ancient saying "none are so blind as those who won't see." Wrongful use of your sensory powers can hamper or prevent progress on a design. Rightful use can focus your attention on the important facts and enable you to block out the less important.

Inhibition can slow the cognitive system outputs. For example, you may "know" which muscles to contract to make a perfect golf swing. Yet you are unable to act on your knowledge. So you produce a terrific hook or slice. As another example, a person may be unable to speak clearly because he can't coordinate the complex muscle workings correctly.

Some people are afflicted with inhibition or malfunction of the data processing function. This may be due to brain damage or psychoses. Thus, damage can cause the symptoms called "aphasia." Such a person can learn (at least in some areas) but at a much slower pace than normal. For example, they may be very slow in learning to read or even unable to learn how to.

Such shortcomings may result from difficulty in storing and retrieving information. In some cases, poor reading ability may be ascribed to the inability to search out the meaning of words or phrases. Psychological problems may cause complete blocks of certain memory areas: some experiences may be so painful that they are "blocked out." Psychiatry may be necessary to uncover these blocks and restore normal mental capacities.

Partial or total inhibition of the complete functioning of the mind may be due to many causes. For example, the mental processes of the brain may be distorted by alcohol, drugs, or a psychosis. In certain fields, distortions may cause unreal images of value. That classic of English literature "The Confessions of an Opium Eater" is one example that comes to mind. Some of the paintings of Salvatore Dali make you wonder about the inner workings of his brain.

Some mental problems caused by inhibitory blocks may be successfully treated. Hypnosis is sometimes used as a tool to raise subconscious thoughts to the level where they may be expressed. Truth drugs provide another useful method.

Anxiety

Anxiety involves both the cognitive and affective systems. But anxiety or fear can impede the operation of any of the functional blocks represented in the cognitive system diagram.

As just one example of such widespread involvements, consider "lying." Now lying is a very complex phenomenon. The liar has one or more emotional responses such as anxiety, fear, or guilt. Because of his emotional response, he lies deliberately and purposefully. Furthermore, he knows he is lying.

Because of the lying he shows physiological effects. There are involuntary changes in salivation, in heart rate, in breathing rate and depth, and in skin temperature.

Besides these physiological effects, he shows emotional tension. He may hesitate, stammer, fidget, or perspire. His face may redden.

To be sure, telling a "white lie" may call forth none of these effects so they can be easily observed. But they are present to some degree. The lie detector (the polygraph) depends on these uncontrollable reactions to tell whether a subject is lying.

Frustration

A designer creates what has never been. So does an artist. So does a composer.

Unfortunately, many attempts at innovation are frustrated. They fail in attainment because of some outside effect.

By definition an attempt to innovate is frustrated when it is made unattainable by any means, with or without design. This definition is important.

Why?

Because in your design work, you may be frustrated by your own inability to carry it to a conclusion. The interfering factor may be some personal problem. Or it may be an unsympathetic superior. In the first instance, you may be unaware of what is blocking you. In the second instance, your superior may be unaware.

Anxieties or fears may be the basis of frustration. The causes of these anxieties or fears may be buried deep within the subconscious.

Aside from frustration, a superior may keep you from doing necessary work or delay you. In dictionary terms, to hinder is to embarrass; to prevent. He may even obstruct you by putting something in the way of progress. He may deny you the use of needed facilities.

Frustration, hindrance, obstruction—these are not just words. They describe real-life situations in the path of a new design. You yourself may cause them. Or your superiors may be the source. To complete a new design, they must either be missing or overcome.

Information-processing rates of the brain

There are limits on how fast you can process input information. Only a relatively modest number of experimenters have worked in this important field. Some of their results are quite surprising—and even unflattering to the human being when he is compared with some kinds of machines and some modern-day computers. For example, the eye can follow up to about twenty cycles per second of lateral movement. The hands can vibrate perhaps ten cycles per second maximum. And when you think about it, this represents a rapid rate of generating musical notes on a piano.

Considered as an overall system, the eye, the cognitive system, and the hand has trouble operating as fast as about a cycle per second.

The capacity of the eye is impressively greater. Some calculations indicate that the eye can transmit information to the optic nerve—if not to the cognitive system itself—at a rate of 4.3 million bits per second. As a comparison, a TV camera can generate more than 15 million bits per second. Microfilm recorders can handle 2.5 million bits per second; some light-sensitive detectors used for input data to computers, 25 thousand bits per second.

Compared with the eye, the ear is able to transmit only 10's of thousands of bits per second to the cognitive system. But you can't react at any such rate.

In one experimental technique, photographs, printed letters and words, and diagrams and drawings are flashed on a screen for very short intervals which may be controlled precisely. The degree of correct recognition of sequences of letters can be measured. The results show that the intake of visual information is at most between 40 and 50 bits per second.

Other experiments have come up with equally surprising results. Where the information signals differ in only one dimension such as frequency or amplitude, the capacity of the human channel is very limited—2 or 3 bits per second. The results are about the same on tests for intensity of tone, visual position of a pointer on a line, and color hues. It seems to be constant for all of your sensory inputs for signals singly and differing in only one dimension. As mentioned earlier, the power of discrimination between two signals presented simultaneously is far greater.

Even more interesting, after looking at or listening to a sequence of letters, numbers, or words, an average person can call back only about seven. The range of usual variation is from five to nine. The limit seems to lie in the amount of short-term storage in the cognitive system.

The highest information transmission rates through the human channel have been obtained with the reading of words: about 40 bits per second. Apparently, the limitation is in information processing—not in the mechanical output rate.

In one set of experiments, discreet word lists were read no faster silently than aloud. Yet very high silent reading rates are widely quoted. However, in reading most books and newspapers, you need not recognize every word to get the sense. Experimenters concluded that the tests indicate the maximum reading rate is determined by mental rather than physical limitations.

In other tests, the same experimenters also showed that just saying words does not determine the observed reading speeds. The test subjects could say a memorized phrase at a far higher rate than they could read unfamiliar words. These tests again suggest that word recognition and not speaking speed accounts for differences of people's reading rates.

A word of caution should be added. Reading tests may not tell the whole story about the amount of information carried by speech waves. For one example, you are able to perform some sort of information processing that enables you to recognize a particular friend's voice, even though you may not have heard it for a long time. How is this done? No one really knows. As a second example, consider the fact that we become aware of shades of meaning in what is being said to us. Information is carried by such diverse means as the tones of the voice, expressions of the face, and even movements of the hands or other parts of the body.

The observed reading rates of 40 to 50 bits per second are far higher than the rate of receiving text in the International Morse code: about 0.58 words per second.

The reading rate for arabic numbers is about the same as for familiar words. Each numeral is an individual pattern to be recognized. It appears that a certain time is required to recognize the pattern regardless of length. This time governs the reading rate up to the point at which words are uttered as fast as possible.

The quoted reading rates of 40 to 50 bits per second are much slower than the capabilities of a telephone or television circuit: 50,000 bits per second for a telephone circuit and 50 million bits per second for a TV circuit. Apparently there is a fundamental limit to the ability of a human being to absorb information. Perhaps a partial explanation is that we can look at any portion of a received TV picture. If the camera and the received picture followed our eye movements, a much less detailed picture would be adequate. Even with the eyes fixed, we can concentrate on a particular part of our visual field.

These limitations of our ability to process information must be taken into account in the design of man-machine systems. It is possible to calculate some reasonable maximum limits of control that cannot be exceeded.

Summary

This chapter is devoted to an example in depth of the system procedure.

As is true of any system design or innovation, it starts with an idea. The idea, chosen from many possibilities, is to draw a model of the human brain.

The approach is to synthesize a functional block diagram, starting with the most general "black box" representation. Step by step, the model is made more detailed and more complex.

An analysis of the model reveals some important facts:

1. Many—and probably most—of the necessary functional blocks are shown. Sufficiency cannot be proved—in fact, the model very likely is incomplete.

2. There are many gaps in our knowledge of the brain. Therefore, it is impossible to tell how—or even where—some necessary functions are implemented.

3. The brain has definite limitations in both memory capabilities and processing. These limitations are extremely important in some aspects of system design. Data are presented on the maximum rates in performing some information-processing tasks.

In conclusion, we give an example of an exploration of a fascinating problem using the system-design procedure. Although the results are not complete, the reason is pointed out: gaps in our knowledge. Many system designs encounter these gaps. They must be filled before further progress is possible. Thus the example accurately points out some of the obstacles to easy innovation.

Bibliography

This book represents a synthesis of ideas gathered by wide reading of books and articles in quite a number of publications in diverse disciplines. In our files a bibliography of some hundreds of items covers the single topic of "Creativity." Rather than go to such extreme lengths in presenting a bibliography, a reasonable number of references to important books is appended. We feel these may be useful to those interested in further reading.

In addition, a list of magazines and learned publications is given. These sometimes publish pertinent material. Unfortunately, to uncover a good idea—or even a provocative idea—an enormous amount of screening must be done. Quite often, the idea is contained in only a sentence or so in a lengthy article. Hence only a modest list is given of papers frequently consulted in writing this book.

Books

Ackoff, R. L., and M. W. Sasieni, *Fundamentals of Operations Research,* Wiley, New York, 1968.

Archibald, R. D., and R. L. Villoria, *Network-Based Management Systems* (PERT/CPM), Wiley, New York, 1967.

Arnold, J. E., "Creativity in Engineering," in *Trans. S.A.E.,* **64,** 17–23 (1956); Quoted by Haefele at length (q.v. below).

Asimow, M., *Introduction to Design,* Prentice-Hall, Englewood Cliffs, New Jersey, 1962.

Becker, J., and R. M. Hayes, *Information Storage, and Retrieval: Tools, Elements, Theories,* Wiley, New York, 1963.

Beer, S., *Decision and Control,* Wiley, New York, 1966.

Blum, G. S., *A Model of the Mind,* Wiley, New York, 1961.

Braithwaite, R. B., *Scientific Investigation,* Cambridge University Press, 1953.

Bruner, J. S., J. J. Goodnow, and G. A. Austin, *A Study of Thinking,* Science Editions, New York, 1962.

Bursk, E. C., and J. F. Chapman, Eds., *New Decision-Making Tools for Engineers, Mathematical Programming as an Aid in the Solving of Business Problems,* Harvard University Press, Cambridge, Mass., 1963.

Cherry, C., *On Human Communication,* Wiley, New York, 1957; 2nd ed. 1968.

Chestnut, H., *Systems Engineering Tools,* Wiley, New York, 1965.

Chestnut, H., *Systems Engineering Methods,* Wiley, New York, 1967.

Churchman, C. W., R. L. Ackoff, and E. L. Arnoff, *Introduction to Operations Research,* Wiley, New York, 1967.

Cofer, C. N., and M. H. Appley, *Motivation: Theory and Research,* Wiley, New York, 1964; Extensive bibliography on psychological aspects of motivation.

Crawford, R. P., *The Techniques of Creative Thinking,* Hawthorn Books, New York, 1954.

Dean, B. V., Ed., *Operations Research in Research and Development,* Wiley, New York, 1963.

Eckman, D. P., Ed., *Systems: Research and Design,* Wiley, New York, 1961.

Eddison, R. T., K. Pennycuick, and B. H. P. Rivett, *Operational Research in Management,* Wiley, New York, 1962.

Felix, L., *The Modern Aspect of Mathematics,* trans. by J. H. Hlavaty and F. C. Hlavaty, Science Editions, New York, 1961.

Flesch, R., *The Art of Clear Thinking,* Harper and Brothers, New York, 1952.

Gosling, W., *The Design of Engineering Systems,* Wiley, New York, 1962.

Guilford, J. P., *The Nature of Human Intelligence,* McGraw-Hill, New York, 1967.

Gutemakher, L. I., *Electronic Information-Logic Machines,* transl. from the Russian by R. Kent, Interscience, New York, 1963.

Haefele, J. W., *Creativity and Innovation,* Reinhold Publ. Co., New York, 1962.

Hare, V. C., Jr., *Systems Analysis: A Diagnostic Approach,* Harcourt Brace and World, New York, 1967.

Harrisberger, L., *Mechanization of Motion. Kinematics-Synthesis Analysis,* Wiley, New York, 1961.

Hutchinson, E. D., *How to Think Creatively,* Abingdon Press, New York, 1949.

Kahn, D., *The Codebreakers, The Story of Secret Writing,* Macmillan, New York, 1963.

Kepner, C. H., and B. B. Tregoe, *The Rational Manager,* McGraw-Hill, New York, 1965; An excellent book on "troubleshooting."

Koestler, A., *The Act of Creation: A Study of the Conscious and Unconscious Processes in Humor, Scientific Discovery, and Art,* Macmillan, New York, 1964.

Lehman, H. C., *Age and Achievement,* Princeton University Press, Princeton, New Jersey, 1953.

Mattson, R. H., *Electronics,* Wiley, New York, 1966.

Meadow, C. T., *The Analysis of Information Systems,* Wiley, New York, 1967.

National Academy of Sciences Report to Committee on Science and Astronautics, U.S. House of Representatives, *Applied Science and Technological Progress,* Government Printing Office, Washington, D.C., 1967.

Osborn, A., *Your Creative Power,* Paperback, Dell Publ. Co., New York, 1948.

Osborn, A. F., *Applied Imagination: Principles and Procedures of Creative Thinking,* Charles Scribner's Sons, New York, 1953.

Paulsen, F. R., Ed., *Contemporary Issues in American Education,* University of Arizona Press, Tucson, 1967; see M. Hughes: *Dimensions of Creativity,* p. 27.

Pelz, D. C., and F. M. Andrews, *Scientists in Organizations,* Wiley, New York, 1966.

Penfield, W., and L. Roberts, *Speech and Brain-Mechanisms,* Princeton University Press, Princeton, New Jersey, 1959.

Piel, G., *Science in the Cause of Man,* Vintage Books, Random House, New York, 1964.

Polya, G., *Mathematics and Plausible Reasoning,* 2 Vols., Princeton University Press, Princeton, New Jersey, 1954.

Polya, G., *How to Solve It,* Princeton University Press, Princeton, New Jersey, 1954; Paperback, Doubleday Anchor Books, Garden City, New York, 1957.

Popper, K. R., *The Logic of Scientific Discovery,* Science Editions, New York, 1961.

Rossman, J., *The Psychology of the Inventor,* Inventor's Publ. Co., Washington, 1931.

Sarton, G., *The History of Science,* 2 Vols., Science Editions, Wiley, New York, 1964.

Schuchman, A., *Scientific Decision Making in Business,* Holt, Rinehart and Winston, New York, 1963.

Shubik, M., Ed., *Game Theory and Related Approaches to Social Behavior,* Wiley, New York, 1964.

Taton, R., *Reason and Chance in Scientific Discovery,* Transl. by A. J. Pomerans, Science Editions, New York, 1962.

Taylor, C. W., Ed., *Widening Horizons in Creativity: The Proceedings of the Fifth Utah Creativity Research Conference,* Wiley, New York, 1964.

Taylor, C. W., and F. Barron, Eds., *Scientific Creativity: Its Recognition and Development,* Wiley, New York, 1963; Selected papers from the first, second, and third University of Utah conferences.

von Neumann, J., *The Computer and the Brain,* Yale University Press, New Haven Connecticut, 1958.

Weil, R., Jr., *The Art of Practical Thinking,* Simon and Schuster, New York, 1940.

Wilson, I. G., and M. E. Wilson, *Information, Computers, and System Design,* Wiley, New York, 1965.

Magazines and Publications

Among the publications which have published useful items are: Arizona Review of the University of Arizona; Behavioral Science; Bell Telephone Laboratories Record; Bell System Technical Journal; Fortune; Industrial Research (many short items of interest); International Science and Technology; Journal of Research and Development of IBM; Management Science (3 series published by the Institute of Management Science—Theory, Applications, and Bulletin); Missiles and Rockets (no longer published); Mental Health Reports of the University of Michigan; Proceedings of the IEEE; Scientific American; Spectrum of the IEEE.

For descriptions of the development of complex systems, The Bell System Technical Journal and IBM Journal are recommended. From time to time, a complete issue is devoted to one system.

Articles

Bronell, A. B., "Is Philosophy off Limits," in *IEEE Spectrum,* 1–5 (May 1964).

DeSimone, D. V., "Education for Innovation," in *IEEE Spectrum,* 5–1 (January 1968).

Dowling, J. E., "Night Blindness," in *Sci. Am.,* 78 (October 1966).

Easton, W. H., "Creative Thinking and How to Develop It," in *Mech. Eng.,* 694–704 (August 1964).

Hartley, R. V. L., "Transmission of Information," in *Bell Sys. Tech. J.,* 7 (July 1928); Measures of Information, units.

Hormann, A., "How a Computer Can Learn," in *IEEE Spectrum,* **1–7** (July 1964).

Hyman, R., "On Prior Information and Creativity," in *Psychol. Reports,* **9,** 151–161 (1961).

Hyman, R., and B. Anderson, "Solving Problems," in *Intern. Sci. Technol.* (September 1965), 35–41.

McCarthy, J., et al., "Information." in *Sci. Am.* (September 1966). Entire issue.

McMillan, B., and D. Slepian, "Information Theory," *Proc. IRE,* **50** (May 1962).

Nievergelt, J. "Computers and Computing—Past, Present, and Future," in *IEEE Spectrum,* **5–1** (January 1968).

Salton, G., "Progress in Automatic Information Retrieval," in *IEEE Spectrum,* **2–8** (August 1965).

Scheerer, M., "Problem-Solving," in *Sci. Am.* (April 1963).

Shannon, C. E., "A Mathematical Theory of Communication," in *Bell Sys. Tech. J.,* **27** (July, October 1948); A classical paper.

Wang, H., "Games, Logic, and Computers," in *Sci. Am.* (November 1965).

Index